Multi-Core Cache Hierarchies

Synthesis Lectures on Computer Architecture

Editor
Mark D. Hill, *University of Wisconsin*

Synthesis Lectures on Computer Architecture publishes 50- to 100-page publications on topics pertaining to the science and art of designing, analyzing, selecting and interconnecting hardware components to create computers that meet functional, performance and cost goals. The scope will largely follow the purview of premier computer architecture conferences, such as ISCA, HPCA, MICRO, and ASPLOS.

Multi-Core Cache Hierarchies

Rajeev Balasubramonian, Norman P. Jouppi, and Naveen Muralimanohar

ISBN: 978-3-031-00606-7   paperback
ISBN: 978-3-031-01734-6   ebook

DOI 10.1007/978-3-031-01734-6

A Publication in the Springer series
SYNTHESIS LECTURES ON COMPUTER ARCHITECTURE

Lecture #17
Series Editor: Mark D. Hill, *University of Wisconsin*
Series ISSN
Synthesis Lectures on Computer Architecture
Print 1935-3235   Electronic 1935-3243

Multi-Core Cache Hierarchies

Rajeev Balasubramonian
University of Utah

Norman P. Jouppi
HP Labs

Naveen Muralimanohar
HP Labs

SYNTHESIS LECTURES ON COMPUTER ARCHITECTURE #17

ABSTRACT

A key determinant of overall system performance and power dissipation is the cache hierarchy since access to off-chip memory consumes many more cycles and energy than on-chip accesses. In addition, multi-core processors are expected to place ever higher bandwidth demands on the memory system. All these issues make it important to avoid off-chip memory access by improving the efficiency of the on-chip cache. Future multi-core processors will have many large cache banks connected by a network and shared by many cores. Hence, many important problems must be solved: cache resources must be allocated across many cores, data must be placed in cache banks that are near the accessing core, and the most important data must be identified for retention. Finally, difficulties in scaling existing technologies require adapting to and exploiting new technology constraints.

The book attempts a synthesis of recent cache research that has focused on innovations for multi-core processors. It is an excellent starting point for early-stage graduate students, researchers, practitioners who wish to understand the landscape of recent cache research. The book is suitable as a reference for advanced computer architecture classes as well as for experienced researchers and VLSI engineers.

KEYWORDS

computer architecture, multi-core processors, cache hierarchies, shared and private caches, non-uniform cache access (NUCA), quality-of-service, cache partitions, replacement policies, memory prefetch, on-chip networks, memory cells.

To our highly supportive families and colleagues.

Contents

Preface

The multi-core revolution is well under-way. The first few mainstream multi-core processors appeared around 2005. Today, it is nearly impossible to buy a desktop or laptop that has just a single core in it. The trend is obvious; the number of cores on a chip will likely double every two or three years. Such processor chips will be widely used in the high-performance computing domain: in supercomputers, servers, and high-end desktops. Just as the volume of low-end devices (for example, smartphones) is expected to increase, the volume of high-end devices (servers in datacenters) is also expected to increase. The latter trend is likely because users will increasingly rely on the *"cloud"* for data storage and computation.

For many decades, one of the key determinants of overall system performance has been the memory hierarchy. Access to off-chip memory consumes many cycles and many units of energy. The more data that can be accommodated and found in the caches of a processor chip, the higher the performance and energy efficiency. This continues to be true in the multi-core era. In fact, multi-core processors are expected to place even higher pressure on the memory system: the number of pins on a chip is expected to remain largely constant while the number of cores that must be fed with data is expected to rise sharply. This makes it even more important to minimize off-chip accesses.

Memory hierarchy efficiency is a strong function of the access latencies of on-chip caches and their hit rates. Future last-level caches (LLCs) are expected to occupy half the processor chip's die area and accommodate many mega-bytes of data. The LLC will likely be composed of many banks scattered across the chip. Access to data will require navigation of long wires and traversal through multiple routing elements. Each access will therefore require many tens of cycles of latency and many nanojoules of energy, depending on the distance that must be traveled. Cache resources will have to be allocated across threads and parts of the LLC may be private to a thread while other parts may be shared by multiple threads.

As a result, caching techniques will undergo evolution in the coming years because of new challenges imposed by multi-core platforms and workloads. Cache policies must now worry about interference among threads as well as large and non-uniform latencies and energy for data transmission between cache banks and cores. On-chip non-local wires continue to scale poorly, increasing the role of the interconnect during cache access. It is therefore imperative that we (i) devise caching policies that reduce long-range communication and (ii) create low-overhead networks to better handle long-range communication when it is required. Several new technology phenomena will also require innovation within the caches. These include the emergence of parameter variation, hard and soft error rates, leakage energy in caches, and thermal constraints from 3D stacking.

Consider the following examples of the game-changing impact of multi-core on caching policies. After years of reliance on LRU-like policies for cache replacement, several papers have emerged in recent years that have shown that alternative approaches are much more effective for replacement in multi-core LLCs. Likewise, the past decade has seen many papers that consider variations of private and shared LLCs, attempting to combine the best of both worlds. Hence, the past and upcoming decades are exciting times for cache research. In retrospect, it should have been obvious that multi-core processors would be imminent; many papers in the 1990s had pointed to this trend. Yet, overall, the community was a little slow to embrace multi-core research. As a result, the pace of multi-core research saw an acceleration only after the arrival of the first commercial multi-core processors. Much work remains, especially for the memory hierarchies of future many-core processors.

Book Organization

The goal of this book is to synthesize much of the recent cache research that has focused on innovations for multi-core processors. For any researcher or practitioner that wishes to understand the landscape of recent cache work, we hope that the book will be an ideal starting point. We also expect early-stage graduate students to benefit from such a synthesis lecture. The book should also serve as a good reference book for advanced computer architecture classes. We expect that the material here will be accessible to both computer scientists and VLSI engineers. The book is not intended as a substitute to reading relevant full papers and chasing down older references. The book will hopefully improve one's breadth and awareness of a multitude of caching topics, while making research on a specific topic more efficient.

Given the vastness of the memory hierarchy topic, we had to set some parameters for what would be worthy of inclusion in this book. We have primarily focused on recent work (2004 and after) that has a strong connection with the use of multiple cores. We have focused our coverage on papers that appear at one of the four primary venues for architecture research: ISCA, MICRO, ASPLOS, and HPCA. However, the book has several discussions of papers that have appeared at other venues and that have made a clear impact within the community. In spite of our best efforts, we have surely left out a few papers that deserve mention; we can hopefully correct some of our oversights in subsequent versions. We encourage readers to contact us to point out our omissions.

The area of multi-core caching has a strong overlap with several other areas within computer architecture. We have explicitly left some of these areas out of this book because they have been covered by other synthesis lectures:

- Off-chip memory systems [10]

- On-chip network designs [11]

- Core memory components (load-store-queue, L1 cache) [12]

- Cache coherence and consistency models [13]

- Phase change memory [14]

- Power optimizations [15]

The discussions in the book have attempted to highlight the key ideas in papers. We have attempted to convey the novelty and the qualitative contribution of each paper. We have typically not summarized the quantitative improvements of each idea. We realize that the mention of specific numbers from papers may be misleading as each paper employs different benchmarks and simulation infrastructure parameters.

The second chapter provides background and a taxonomy for multi-core cache hierarchies. The third chapter then examines policies that bridge the gap between shared LLCs and private LLCs. The papers in Chapter 2 typically assume that the LLC is made up of a collection of banks, with varying latencies to reach each bank. The considered policies attempt to place data in these banks so that access latency is minimal and hit rates can be maximized. Chapter 3 also focuses on hit rate optimization, but it does so for a single cache bank. Instead of moving data between banks, the considered policies improve hit rates with better replacement policies, better organizations for associativity, and block-level optimizations (prefetch, dead block prediction, compression, etc.). That chapter also examines how a single cache bank can be partitioned among multiple threads for high throughput and quality-of-service. Since access to an LLC often requires navigation of an on-chip network, Chapter 4 describes on-chip network innovations that have a strong interaction with caching policies. Chapter 5 describes how modern technology trends are likely to impact the design of future caches. It covers modern technology phenomena such as 3D die-stacking, parameter variation, rising error rates, and emerging non-volatile memories. Chapter 6 concludes with some thoughts on avenues for future work.

Figure 1 uses the same classification as above and shows the number of papers that have appeared in each cache topic in the past seven years at the top four architecture conferences. Note that there can be multiple ways to classify the topic of a paper, and the data should be viewed as being approximate. The data serves as an indicator of hot topics within multi-core caching. Activity appears to be highest in technology phenomena, block prefetch, and shared caches.

Rajeev Balasubramonian, Norman P. Jouppi, and Naveen Muralimanohar
May 2011

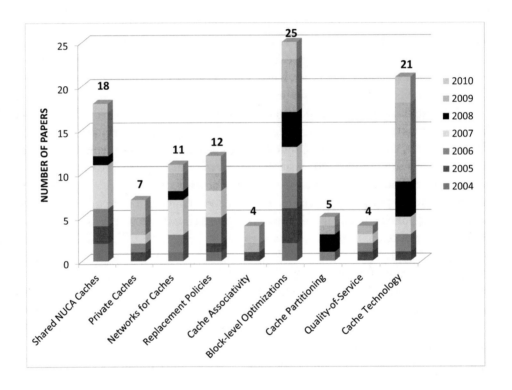

Figure 1: Number of papers in various cache topics in the last seven years at ISCA, MICRO, ASPLOS, and HPCA.

Acknowledgments

We thank everyone that provided comments and feedback on early drafts of this synthesis lecture, notably, Mark Hill, Aamer Jaleel, Gabriel Loh, Mike Morgan, and students in the Utah Arch research group.

Rajeev Balasubramonian, Norman P. Jouppi, and Naveen Muralimanohar
May 2011

CHAPTER 1

Basic Elements of Large Cache Design

This chapter presents a landscape of cache hierarchy implementations commonly employed in research/development and identifies their key distinguishing features. These features include the following: shared vs. private, centralized vs. distributed, and uniform vs. non-uniform access. There is little consensus in the community about what constitutes an optimal cache hierarchy implementation. Some levels of the cache hierarchy employ private and uniform access caches, while other levels employ shared and non-uniform access. We will point out the pros and cons of selecting each feature, and it is perfectly reasonable for a research effort to pick any combination of features for their baseline implementation. Much of the focus of this book is on the design of the on-chip *Last-Level Cache (LLC)*. In the past, most on-chip cache hierarchies have been comprised of two levels (L1 and L2), but it is becoming increasingly common to incorporate three levels in the cache hierarchy (L1, L2, and an L3 LLC). As we explain in this chapter and the next, future LLCs have a better chance of optimizing miss rates, latency, and complexity if they are implemented as shared caches. This chapter also discusses other basics that are required to understand modern cache innovations.

Before getting started, a couple of terminology clarifications are in order. In a cache hierarchy, a cache level close to the processor is considered an "upper-level" cache, while a cache level close to main memory is considered a "lower-level" cache. We will also interchangeably use the terms "cache line" and "cache block", both intended to represent the smallest unit of data handled during cache fetch and replacement.

1.1 SHARED VS. PRIVATE CACHES

Most modern high-performance processors incorporate multiple levels of the cache hierarchy within a single chip. In a multi-core processor, each core typically has its own private L1 data and L1 instruction caches. Considering that every core must access the L1 caches in nearly every cycle, it is not typical to have a single L1 cache (either data or instruction cache) shared by multiple cores. A miss in the L1 cache initiates a request to the L2 cache. For most of the discussion, we will assume that the L2 is the LLC. But the same arguments will also apply to an L3 LLC in a 3-level hierarchy, where the L1 and L2 are private to each core. We first compare the properties of shared and private LLCs.

1.1.1   SHARED LLC

A single large L2 LLC may be shared by multiple cores on a chip. Since the requests originating from a core are filtered by its L1 caches, it is possible for a single-ported L2 cache to support the needs of many cores. One example organization is shown in Figure 1.1. In this design, eight cores

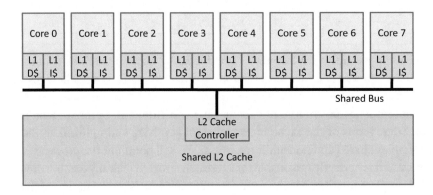

Figure 1.1: Multi-core cache organization with a large shared L2 cache and private L1 caches per core. A snooping-based cache coherence protocol is implemented with a bus connecting the L1s and L2.

share a single large L2 cache. When a core fails to find data in its L1 caches, it places the request on a bus shared by all cores. The L2 cache controller picks requests off this bus and performs the necessary look-up. In such a large shared L2 cache, there are no duplicate copies of a memory block, but a given block may be cached in multiple different L1 caches. Coherence must be maintained among the L1s and the L2. In the bus-based example in Figure 1.1, coherence is maintained with a snooping-based protocol. Assume that the L1 caches employ a write-back policy. When a core places a request on the bus, every other core sees this request and looks up its L1 cache to see if it has a copy of the requested block. If a core has a copy of the block in modified state, *i.e.*, this copy happens to be the most up-to-date version and the only valid copy of the block, the core must respond by placing the requested data on the bus. If no core has the block in modified state, the L2 cache must provide the requested data. The L2 cache controller figures out that it must respond by examining a set of control signals that indicate that the cores have completed their snoops and do not have the block in modified state. If the requesting core is performing a write, copies of that block in other L1 caches are invalidated during the snoop operation. Of course, there can be many variations of this basic snooping-based protocol [16, 17]. If a write-through policy is employed for the L1 caches, an L1 miss is always serviced by the L2. A write-through policy can result in significant bus traffic and energy dissipation; this overhead is not worthwhile in the common case. Similarly, write-update cache coherence protocols are also more traffic intensive and not in common use. However, some of these design guidelines are worth re-visiting in the context of modern single-chip multi-cores with relatively cheap interconnects.

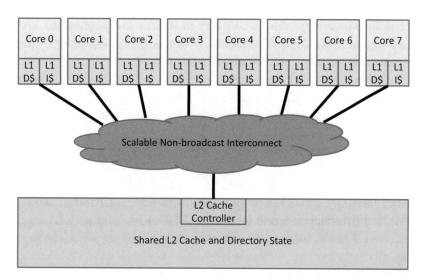

Figure 1.2: Multi-core cache organization with a large shared L2 cache and private L1 caches per core. A scalable network connects the L1 caches and L2 and a directory-based cache coherence protocol is employed. Each block in the L2 cache maintains directory state to keep track of copies cached in L1.

The above example primarily illustrates the interface required before accessing a shared cache. In essence, a mechanism is required to ensure coherence between the shared L2 and multiple private L1s. If the number of cores sharing an L2 is relatively small (16 or fewer), a shared bus and a snooping-based coherence protocol will likely work well. For larger-scale systems, a scalable interconnect and a directory-based coherence protocol are typically employed. As shown in Figure 1.2, the cores and the L2 cache are connected with some scalable network and broadcasting a request is no longer an option. The core sends its request to the L2 cache and each L2 block is associated with a directory that keeps track of whether other L1 caches have valid copies of that block. If necessary, other caches are individually contacted to either invalidate data or obtain the latest copy of data.

There are many advantages to employing a shared cache. First, the available storage space can be dynamically allocated among multiple cores, leading to better utilization of the overall cache space. Second, if data is shared by multiple cores, only a single copy is maintained in L2, again leading to better space utilization and better cache hit rates. Third, if data is shared by multiple cores and subject to many coherence misses, the cache hierarchy must be navigated until the coherence interface and shared cache is encountered. The sooner a shared cache is encountered, the sooner coherence misses can be resolved.

The primary disadvantages of a shared cache are as follows. The working sets of different cores may interfere with each other and impact each other's miss rates, possibly leading to poorer quality-of-service. As explained above, access to a shared L2 requires navigation of the coherence interface: this may impose overheads if the cores are mostly dealing with data that is not shared by multiple cores. Finally, many papers cite that a single large shared L2 cache may have a relatively long access time on average. Also, a core may experience many contention cycles when attempting to access a resource shared by multiple cores. However, we will subsequently (Section 2.1) show that both of these disadvantages can be easily alleviated.

It must be noted that many of our examples assume that the L2 cache is inclusive, *i.e.*, if a data block is present in L1, it is necessarily also present in L2. In Section 1.4, we will discuss considerations in selecting inclusive and non-inclusive implementations.

1.1.2   PRIVATE LLC

A popular alternative to the single shared LLC is a collection of private last-level caches. Assuming a two-level hierarchy, a core is now associated with private L1 instruction and data caches *and* a private unified (handling data and instructions) L2 cache. A miss in L1 triggers a look-up of the core's private L2 cache. Some of the advantages/disadvantages of such an organization are already apparent.

The working sets of threads executing on different cores will not cause interference in each other's L2 cache. Each private L2 cache is relatively small (relative to a single L2 cache that must be shared by multiple cores), allowing smaller access times on average for L2 hits. The private L2 cache can be accessed without navigating the coherence interface and without competition for a shared resource, leading to performance benefits for threads that primarily deal with non-shared data.

A primary disadvantage of private L2 caches is that a data block shared by multiple threads will be replicated in each thread's private L2 cache. This replication of data blocks leads to a lower effective combined L2 cache capacity, relative to a shared L2 cache of similar total area. In other words, four private 256 KB L2 caches will accommodate less than 1 MB worth of data because of duplicate copies of a block, while a 1 MB shared L2 cache can indeed accommodate 1 MB worth of data. Another disadvantage of a private L2 cache organization is the static allocation of L2 cache space among cores. In the above example, each core is allocated a 256 KB private L2 cache even though some cores may require more or less. In a 1 MB shared L2 cache, it is possible for one core to usurp (say) 512 KB of the total space if it has a much larger working set size than threads on the other cores.

By employing private L2 caches, the coherence interface is pushed down to a lower level of the cache hierarchy. First, consider a small-scale multi-core machine that employs a bus-based snooping coherence protocol. On an L2 miss, the request is broadcast on the bus. Other private L2 caches perform snoop operations and place their responses on the bus. If it is determined that none of the other private L2 caches can respond, a controller forwards this request to the next level of the hierarchy (either an L3 cache or main memory). When accessing shared data, such a private L2 organization imposes greater latency overheads than a model with private L1s and a shared L2. The key differentiating overheads are the following: (i) the private L2 cache is looked up before placing the request on the bus, (ii) snoops take longer as a larger set of tags must be searched, and (iii) it takes longer to read data out of another large private L2 data array (than another small private L1 cache).

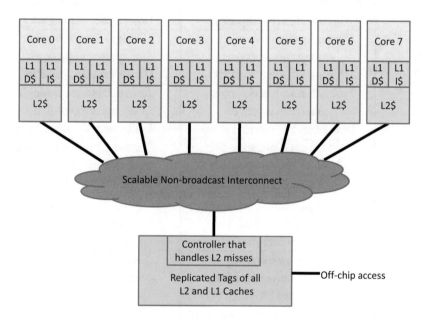

Figure 1.3: Multi-core cache organization where each core has a private L2 cache and coherence is maintained among the private L2 caches with a directory-based protocol across a scalable non-broadcast interconnect.

The coherence interface is even more complex if a directory-based protocol is employed (shown in Figure 1.3). On an L2 miss, the request cannot be broadcast to all cores, but it is sent to a directory. This directory may be centralized or distributed, but in either case, long on-chip distances may have to be traversed. This directory must keep track of all blocks that are cached on chip and it essentially replicates the tags of all the private L2 caches. A highly-associative search is required to detect if the requested address is in any of the private L2 caches. If each of the four 256 KB private L2 caches is 4-way set-associative, the directory look-up will require 16 tag comparisons to determine the state of the block. L1 tags need not be replicated by preserving inclusion between the L1s and L2. If a block is detected in another private L2 cache, messages are exchanged between the directory and cores to move the latest copy of data to the requesting core's private cache. On the other hand, the shared L2 cache simply associates the directory with the unique copy of the block in

L2, thus eliminating the need to replicate L2 tags. In short, assuming inclusion, the use of a shared on-chip LLC makes it easier to detect if a cached copy exists on the chip.

Table 1.1: A comparison of the advantages and disadvantages of private and shared cache organizations.

Shared L2 Cache	Private L2 Caches
No replication of shared blocks (higher effective capacity)	Replication of shared blocks (lower effective capacity)
Dynamic allocation of space among threads/cores (higher effective capacity)	Design-time allocation of space among cores (lower effective capacity)
Quick traversal through coherence interface (low latency for shared data)	Slower traversal through coherence interface (high latency for shared data)
No L2 tag replication for directory implementation (low area requirements)	Directory implementation requires replicated L2 tags (high area requirements)
Higher interference between threads (negatively impacts QoS)	No interference between threads (positively impacts QoS)
Longer wire traversals on average to detect an L2 hit (high average hit latency[1])	Short wire traversals on average to detect an L2 hit (low average hit latency)
High contention when accessing shared resource (bus and L2) (high hit latency for private data)	No contention when accessing L2 cache (low hit latency for private data)

The differences between private and shared L2 cache organizations are summarized in Table 1.1. It is also worth noting that future processors may employ combinations of private and shared caches. For example (Figure 1.4), in a 16-core processor, each cluster of four cores may share an L2 cache, and there are four such L2 caches that are each private to their cluster of four cores. Snooping-based coherence is first maintained among the four L1 data caches and L2 cache in one cluster; snooping-based coherence is again maintained among the four private L2 caches.

1.1.3   WORKLOAD ANALYSIS

Some recent papers have focused on analyzing the impact of baseline shared and private LLCs on various multi-threaded workloads. The work of Jaleel et al. [18] characterizes the behavior of bioinformatics workloads. They show that more than half the cache blocks are shared, and a vast majority of LLC accesses are to these shared blocks. Given this behavior, a shared LLC is a clear

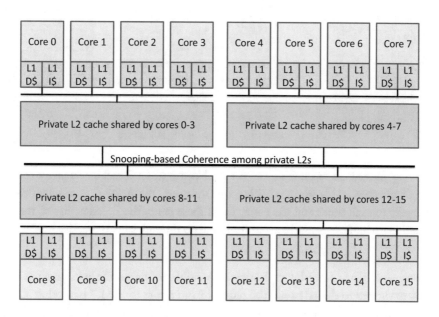

Figure 1.4: 16-core machine where each cluster of four cores has a private L2 cache that is shared by its four cores. There are two hierarchical coherence interfaces here: one among the L1 data caches and L2 cache within a cluster and one among the four private L2 caches.

winner over an LLC that is composed of many private LLCs. Bienia et al. [19] show a workload analysis of the SPLASH-2 and PARSEC benchmark suites, including cache miss rates and the extent of data sharing among threads. Many other cache papers also report workload characterizations in their analysis, most notably, the work of Beckmann et al. [20] and Hardavellas et al. [3].

1.2   CENTRALIZED VS. DISTRIBUTED SHARED CACHES

This section primarily discusses different implementations for a shared last level cache. At the end of the section, we explain how some of these principles also apply to a collection of private caches.

We have already considered two shared L2 cache organizations in Figures 1.1 and 1.2. In both of these examples, the L2 cache and its controller are represented as a single centralized entity. When an L1 miss is generated, the centralized L2 cache controller receives this request either from the bus (Figure 1.1) or from a link on the scalable network (Figure 1.2). It then proceeds to locate the corresponding block within the L2 cache structure. If the L2 cache is large (as is usually the case), it is itself partitioned into numerous banks, and some sort of interconnection network must be navigated to access data within one of the banks (more details on this in later sections). Thus, some form of network fabric may have to be navigated to simply reach the centralized L2 cache controller and yet another fabric is navigated to reach the appropriate bank within the L2 cache. Depending

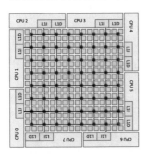 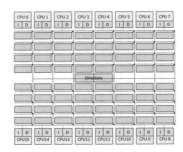 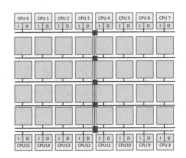

Figure 1.5: Shared L2 cache with a centralized layout (the L2 cache occupies a contiguous area in the middle of the chip and is surrounded by cores). An on-chip network is required to connect the many cache banks to each other and the cores. In all these cases, the functionality of the cache controller is replicated in each of the banks to avoid having to go through a central entity.

on the types of fabrics employed, it may be possible to merge the two. In such a scenario, a single fabric is navigated to directly send the request from a core to the L2 bank that stores the block. The L2 bank will now need some logic to take care of the necessary coherence operations; in other words, the functionality of the L2 cache controller is replicated in each of the banks to eliminate having to go through a single centralized L2 cache controller.

Some example physical layouts of such centralized shared L2 caches are shown in Figure 1.5. Each of these layouts has been employed in research evaluations (for example, [2, 7, 21, 22]). Even though the cache is banked and the controller functionality is distributed across the banks, we will refer to these designs as *Centralized* because the LLC occupies a contiguous area on the chip. Such centralized cache structures attempt to provide a central pool of data that may be quickly and efficiently accessed by cores surrounding it. By keeping the cache banks in close proximity to each other, movement of data between banks (if required) is simplified. The interconnects required between L2 cache banks and the next level of the hierarchy (say, the on-chip memory controller) are simplified by aggregating all the cache banks together. The L2 cache also ends up being a centralized structure if it is implemented on a separate die that is part of a 3D-stacked chip (assuming a single die-to-die bus that communicates requests and responses between CPUs and a single L2 cache controller).

An obvious extension to this model is the *distributed shared L2 cache*. Even though the L2 cache is logically a shared resource, it may be physically distributed on chip, such that one bank of the L2 may be placed in close proximity to each core. The core, its L1 caches, and one bank (or *slice*) of the L2 cache together constitute one *tile*. A single on-chip network is used to connect all the tiles. When a core has an L1 miss, its request is routed via the on-chip network to the tile that is expected to have the block in its L2 bank. Such a tiled and distributed cache organization is desireable because it allows manufacturers to design a single tile and instantiate as many tiles as allowed by the area budget. It therefore lends itself better to scalable design/verification cost, easy

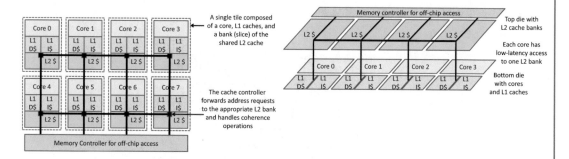

Figure 1.6: A shared L2 cache with a physically distributed layout. One bank (or *"slice"*) of L2 cache is associated with each core. A *"tile"* is composed of a core, its L1 caches, and its associated L2 bank. The figure on the right physically separates the L2 banks onto a separate die in a 3D stack, but it retains the same logical organization as the figure on the left.

manufacture of families of processors with varying numbers of tiles, and simple upgrades to new technology generations. Such distributed shared caches have also been implemented in recent Tilera multi-core processors. Two examples of such a physical layout are shown in Figure 1.6. We will also subsequently see how architectural mechanisms for data placement can take advantage of such a physical organization. The primary disadvantage of this organization is the higher cost in moving data/requests between L2 cache banks and the next level of the memory hierarchy.

While a centralized L2 cache structure may be a reasonable design choice for a processor with a medium number of cores, it may prove inefficient for a many-core processor. Thermal and interconnect (scalability and wire-length) limitations may prevent many cores from surrounding a single large centralized L2 cache. It is therefore highly likely that many-core processors will employ a distributed L2 cache where every core at least has very quick access to one bank of the shared L2 cache. Distributing the L2 cache also has favorable implications for power density and thermals.

Our definitions of *Centralized* and *Distributed* caches are only meant to serve as an informal guideline when reasoning about the properties of caches. A centralized cache is defined as a cache that occupies a contiguous area on the chip. A distributed cache is defined as a cache where each bank is tightly coupled to a core or collection of cores. In Figure 1.6(a), if the layout of cores 4-7 was a mirror image of the cores 0-3, the cache would be classified as both centralized and distributed, which is admittedly odd.

When implementing a private L2 cache organization, it makes little sense to place the core's private L2 cache anywhere but in close proximity to the core. Hence, a private L2 cache organization has a physical layout that closely resembles that of the distributed shared L2 cache just described. In other words, for both organizations, a tile includes a bank of L2 cache; it is the logical policies for placing and managing data within the many L2 banks that determines if the L2 cache space is shared or private.

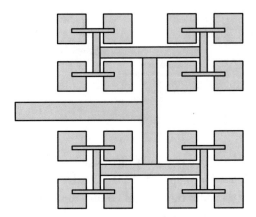

Figure 1.7: An example H-tree network for a uniform cache access (UCA) architecture with 16 arrays.

For the private L2 cache organization, if a core has a miss in its private L2 cache, the request must be forwarded to the coherence interface. If we assume directory-based coherence, the request is forwarded to the directory structure that could itself be centralized or distributed. The considerations in implementing a centralized/distributed on-chip directory are very similar to those in implementing a centralized/distributed shared L2 cache. The on-chip network is therefore employed primarily to deal with coherence operations and accesses to the next level of the hierarchy. The on-chip network for the distributed shared L2 cache is also heavily employed for servicing L2 hits.

1.3   NON-UNIFORM CACHE ACCESS

In most processors until very recently, a cache structure is designed to have a uniform access time regardless of the block being accessed. In other words, the delay for the cache access is set to be the worst-case delay for any block. Such *Uniform Cache Access (UCA)* architectures certainly simplify any associated instruction scheduling logic, especially if the core pipeline must be aware of cache hit latency. However, as caches become larger and get fragmented into numerous banks, there is a clear inefficiency in requiring that every cache access incur the delay penalty of accessing the furthest bank. Simple innovations to the network fabric can allow a cache to support non-uniform access times; in fact, many of the banked cache organizations that we have discussed so far in this chapter are examples of *Non-Uniform Cache Access (NUCA)* architectures.

A UCA banked cache design often adopts an H-tree topology for the interconnect fabric connecting the banks to the cache controller. With an H-tree topology (example shown in Figure 1.7), every bank is equidistant from the cache controller, thus enabling uniform access times to every bank. In their seminal paper [1], Kim et al. describe the innovations required to support a NUCA architecture and quantify its performance benefits. Firstly, variable access times for the L2 are acceptable as the core pipeline does not schedule instructions based on expected L2 access time. Secondly, instead of adopting an H-tree topology, a grid topology is employed to connect banks to the cache controller. The latency for a bank is a function of its size and the number of network hops required to route the request/data between the bank and the cache controller. It is worth noting that messages on this grid network can have a somewhat irregular pattern, requiring complex mechanisms at every hop to support routing and flow control. These mechanisms were not required in the H-tree network, where requests simply radiated away from the cache controller in a pipelined fashion. The complexity in the network is the single biggest price being paid by a NUCA architecture to provide low-latency access to a fraction of cached data. It can be argued that most future architectures will anyway require complex on-chip networks to handle somewhat arbitrary messaging between the numerous cores and cache banks. Especially in tiled architectures such as the ones shown in Figure 1.6, it is inevitable that the different cache banks incur variable access latencies as a function of network distance.

In the next chapter, we will discuss innovations to NUCA architectures that attempt to cleverly place data in an "optimal" cache bank. These innovations further drive home the point that traffic patterns are somewhat arbitrary and require complex routing/flow control mechanisms. In Chapter 4, we describe on-chip network innovations that are applicable to specific forms of NUCA designs. In the rest of this section, we describe the basic NUCA designs put forth by Kim et al. in their paper [1].

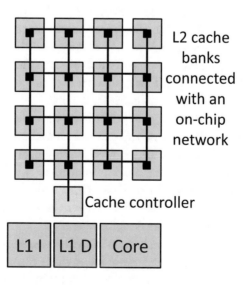

Figure 1.8: A NUCA L2 cache connected to a single core [1].

Physical Design

Kim et al. [1] consider a large L2 cache that has a single cache controller feeding one processor core (see Figure 1.8). In terms of physical layout, they consider two implementations. The first employs a dedicated channel between each of the many cache banks and the cache controller. While private channels can provide low contention and low routing overheads for each access, the high metal area requirements of these private channels are prohibitive. Such a design would also not easily scale to multiple cache controllers or cores. The second layout employs a packet-switched on-chip network with a grid topology. This ensures a tolerable metal area requirement while still providing high bandwidth and relatively low contention. Multiple cache controllers (cores) can be easily supported by linking them to routers on the periphery of the grid (or any router for that matter). While Kim et al. [1] advocate the use of *"lightweight"* routers that support 1-cycle hops, it is not yet clear if such routers can be designed while efficiently supporting the flow control needs of the cache network (more on routers in Chapter 4). If such lightweight routers exist, Kim et al. correctly point out that a highly-banked cache structure is desireable: it reduces access time within a bank, reduces contention at banks and routers, supports higher overall bandwidth, and provides finer-grain control of cache resources.

Logical Policies

Logical policies for data management must address the following three issues: (1) *Mapping:* the possible locations for a data block, (2) *Search:* the mechanisms required to locate a data block, and (3) *Movement:* the mechanisms required to change a block's location.

The simplest mapping policy distributes the sets of the cache across banks while co-locating all ways of a set in one bank. As a result, the block address and its corresponding cache index bits are enough to locate the unique bank that houses that set. The request is routed to that bank, tag comparison is performed for all ways in that set, and the appropriate data block is returned to the cache controller. Since the mapping of a data block to a bank is unique, such an architecture is known as *Static-NUCA* or *S-NUCA*. This design does not support movement of a block between banks and does not require mechanisms to search for a block.

An alternative mapping policy distributes ways and sets across banks. The W ways of a set can be distributed across W different banks. Policies must be defined to determine where a block is placed upon fetch. Similarly, policies are required to move data blocks between ways in order to minimize average access times. Because a block is allowed to move between banks, such an architecture is referred to as *Dynamic-NUCA* or *D-NUCA*. Finally, a search mechanism is required to quickly locate a block that may be in one of W different banks. Kim et al. consider several policies for each of these and the salient ones are described here.

The search of a block can happen in an *incremental* manner, *i.e.*, the closest or most likely bank is first looked up, and if the block is not found there, the next likely bank is looked up. Alternatively, a *multicast* search operation can be carried out where the request is sent to all candidate banks, and they are searched simultaneously. The second approach will yield higher performance (unless it introduces an inordinate amount of network contention) but also higher power. Combinations of the two are possible where, for example, multicast happens over the most likely banks, followed by an incremental search over the remaining banks. Kim et al. propose a *Smart Search* mechanism where a partial tag (six bits) for each block is stored at the cache controller. A look-up of this partial tag structure helps identify a small subset of banks that likely have the requested data and only those banks must now be searched. While such an approach is very effective for a single-core design where all block replacements happen via the cache controller, this design does not scale as well to multi-core designs where partial tags must be redundantly maintained at potentially many cache controllers. The next chapter discusses alternative solutions proposed by Chishti et al. [23, 24]. To date, efficient search in a D-NUCA cache remains an open problem.

To allow frequently accessed blocks to migrate to banks closer to the cache controller, Kim et al. employ a *Generational Promotion* mechanism (and also consider several alternatives). On a cache miss, the fetched block is placed in a way in the furthest bank. Upon every subsequent hit, the block swaps locations with the block that resides in the adjacent bank (way) and edges closer to the cache controller.

Summary

Kim et al. show that NUCA caches are a clear winner over similar sized UCA caches and multi-level hierarchies (although, note that an N-way D-NUCA cache is similar in behavior to a non-inclusive N-level cache hierarchy). D-NUCA policies offer a performance benefit in the 10% range over S-NUCA and this benefit grows to 17% if a smart search mechanism is incorporated. This argues for the use of "clever" data placement, but the feasibility of D-NUCA mechanisms is somewhat questionable. Data movement is inherently complex; consider the example where a block is being searched for while it is in the process of migrating and the mechanisms that must be incorporated to handle such corner cases. The feasibility of smart search, especially in the multi-core domain, is a major challenge. While the performance of D-NUCA is attractive and has sparked much research, recent work shows that it may be possible to design cache architectures that combine the hardware simplicity of S-NUCA and the high performance of D-NUCA. Chapter 2 will discuss several of these related bodies of work.

1.4   INCLUSION

For much of the discussions in this book, we will assume inclusive cache hierarchies because they are easier to reason about. However, many research evaluations and commercial processors employ non-inclusive hierarchies as well. It is worth understanding the implications of this important design choice. Unfortunately, research papers (including those by the authors of this book) often neglect to mention assumptions on inclusion. Section 3.2.1 discusses a recent paper [25] that evaluates a few considerations in defining the inclusion policy.

If the L1-L2 hierarchy is inclusive, it means that every block in L1 has a back-up copy in L2. The following policy ensures inclusion: when a block is evicted from L2, its copy in the L1 cache is also evicted. If a single L2 cache is shared by multiple L1 caches, the copies in all L1s are evicted. This is an operation very similar to L1 block invalidations in a cache coherence protocol.

The primary advantage of an inclusive hierarchy in a multi-core is the ease in locating a data block upon an L1 miss – either a copy of the block will be found in L2, or the L2 will point to a modified version of the block in some L1, or an L2 miss will indicate that the request can be directly sent to the next level of the hierarchy. The L2 cache is also a central point when handling coherence requests from lower levels of the hierarchy (L3 or off-chip). The disadvantage of an inclusive hierarchy is the wasted space because most L1 blocks have redundant copies in L2.

Some L1-L2 hierarchies are designed to be exclusive (a data block will be found in either an L1 cache or the L2 cache, but not in both) or non-inclusive (there is no guarantee that an L1 block has a back-up copy in L2). Data block search is more complex in this setting: on an L1 miss, other L1 caches and the L2 will have to be looked up. If a snooping-based coherence protocol is employed between the L1s and L2, this is not a major overhead as a broadcast and search happens over all L1s on every L1 miss. This search of L1s must be done even when handling coherence requests from lower levels of the hierarchy. However, just as with snooping-based protocols, the search operation does not scale well as the number of L1 caches is increased. The advantage, of course, is the higher overall cache capacity because there is little (or no) duplication of blocks.

Another basic implementation choice is the use of write-through or write-back policies. Either is trivially compatible with an inclusive L1-L2 hierarchy, with write-through policies yielding higher performance if supported by sufficiently high interconnect bandwidth and power budget. A write-through policy ensures that shared blocks can be quickly found in the L2 cache without having to look in the L1 cache of another core. A writeback cache is typically appropriate for a non-inclusive hierarchy.

CHAPTER 2

Organizing Data in CMP Last Level Caches

Multi-cores will likely accommodate many mega-bytes of data in their last-level on-chip cache. As was discussed in Chapter 1, the last-level cache (LLC) can be logically shared by many cores and be either physically distributed or physically contiguous on chip. Alternatively, each core can maintain a private LLC. We will next examine several architectural innovations that attempt to cleverly place data blocks within the LLC to optimize metrics such as miss rates, access times, quality-of-service, and throughput. This chapter primarily focuses on cache architectures that have non-uniform access for different cache banks and techniques that attempt to minimize access times for blocks (we consider both shared and private LLCs in this chapter). The next chapter focuses on data placement techniques that are oblivious or agnostic to the non-uniform nature of cache accesses.

2.1 DATA MANAGEMENT FOR A LARGE SHARED NUCA CACHE

We begin by examining an LLC that is shared by all on-chip cores. It is expected that such a large multi-megabyte cache will offer non-uniform latencies to banks, regardless of whether the LLC cache is physically contiguous or physically distributed across the chip. Most innovations will apply to either physical layout, and in the subsequent discussions, we make note of where this is not true.

In Section 1.3, we have already discussed the basics of a NUCA implementation [1], where banks are connected with an on-chip network and cache access latencies are determined by network distances and contention for each request. We also discussed the basic mechanisms required for data mapping, movement, and search. Note that the work by Kim et al. [1] dealt exclusively with a NUCA design for a single-core processor. The papers discussed in this section attempt to develop solutions for data management in a NUCA cache that support multiple cores. We first discuss a few papers that are characterized by their use of complex search mechanisms but that allow great flexibility in terms of data placement and migration. We then discuss some limited work on LLC data replication. We finally end with a discussion of papers that influence data placement with schemes that do not require complex searches.

2.1.1 PLACEMENT/MIGRATION/SEARCH POLICIES FOR D-NUCA

The work of Kim et al. [1] introduced two major forms of NUCA caches: static (S) and dynamic (D) NUCA. Since D-NUCA was expected to perform better, much of the early NUCA work focused on D-NUCA and allowed blocks to migrate within the LLC. This sub-section focuses almost exclusively on such D-NUCA designs. However, all of this work had to suffer from the overheads of a fairly complex search mechanism, a problem that to date does not have a compelling solution.

Beckmann and Wood, MICRO'04

Beckmann and Wood [2] proposed the first detailed multi-core NUCA architecture. They assume a layout (shown in Figure 2.1) where the shared NUCA cache resides in the middle of the chip and is surrounded by eight cores. The shift to multi-core motivates the following basic changes to the NUCA architecture of Kim et al.: 1. Requests are injected into the NUCA network from eight distributed locations on chip (this results in a much less regular traffic pattern). 2. The cache is broken into even more banks (as many as 256 banks) to support higher bandwidth requirements (later studies [7] have shown that such excessive banking is not required). 3. The on-chip network overheads are alleviated by assuming a larger link width and by connecting each router to four banks.

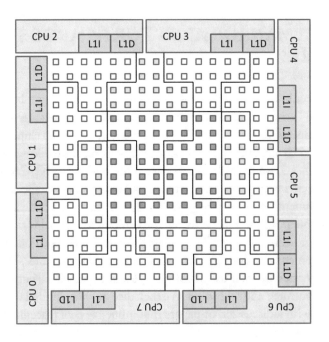

Figure 2.1: A NUCA L2 cache in an 8-core processor [2].

A major contribution of this work is the classification of banks into regions and architectural policies to allow a block to migrate to a region that minimizes overall access times. As shown in Figures 1.5a and 2.1, the 256 cache banks are organized as 16 *tetris*-shaped regions or *bankclusters*. The cache is organized as 16-way set-associative and each region accommodates one way for all sets. Thus, a given block may reside in any one of the 16 regions, depending on the way it gets placed in. Eight of the 16 regions are classified as *local* because of their physical proximity to the corresponding eight cores (these are the lightly shaded regions in the figure). Ideally, a core must cache its private data in its local region. Four *center* regions (dark shade) are expected to cache data that is shared by multiple cores. Four *inter* regions are expected to cache data that is shared by two neighboring cores.

The rules for block placement (initial allocation) and movement are simple. Initial placement is somewhat random (based on the block's tag bits). From here, a block is allowed to gradually migrate to different regions based on the cores that access it. Over time, as cores "pull" the block closer to their local regions, the block eventually settles at the "center-of-gravity" of all its requests. The following migration rules allow a block to move without having to maintain state with each block. "Other-local" refers to the local bank of a core other than the requesting core.

```
other-local => other-inter => other-center => my-center => my-inter => my-local
```

Beckmann and Wood correctly identify that the most significant problem with the above architecture is the difficulty in locating a block. In a dynamic-NUCA policy that distributes ways across banks, the way (bank) that contains the data is not known beforehand. Beckmann and Wood employ a multicast search policy where the block address is first multicast to the six banks most likely to contain the requested block (the local, inter, and four center regions). If all six banks return a miss, then the request is multicast to the remaining 10 banks. While this search mechanism is reasonably fast, it imposes significant network and bank load. Performing the search sequentially across all 16 banks is likely too time-consuming. Kim et al. [1] employed a partial tag structure that allowed a single core to quickly determine the bank at a modest overhead. Extending that solution to a multi-core platform would incur much higher storage overhead and network overhead to keep all partial tag structures coherent after every migration. None of the above search mechanisms (including other innovations discussed at the end of this section) appear acceptable in the modern power- and wire-constrained world. Further, any search mechanism must also be cognizant of the fact that a block may be in transit between banks, else a miss may be signaled even though the block does exist in cache. This requires a mechanism to detect an inconsistency (for example, the directory believes that a cached copy exists, but the search failed to detect it) and then re-execute search after all outstanding migrations have finished. Thus, the lack of a robust search mechanism clearly appears to be the Achilles heel of any multi-core dynamic-NUCA policy that distributes ways across banks. To estimate the impact of search, Beckmann and Wood also evaluate an idealized perfect search mechanism that magically sends the request to the correct bank.

Their results can be best summarized as follows. Compared to a static-NUCA policy that requires no search but does not strive for data proximity, the dynamic-NUCA policy with the multicast search offers almost no overall benefit. This is because the benefits of block migration (proximity of data) are offset by the overheads of an expensive search mechanism. An idealized perfect search is required to extract a few percentage points of improvement. Even with idealized search, improvements are hard to come by because many shared blocks tend to reside in central banks and cannot be accessed with very low latencies[1]. Beckmann and Wood show that prefetching data into L1 with stream buffers is perhaps a more practical way to minimize the performance impact of long access times in a large shared NUCA cache (although, note that prefetch will exacerbate the network power problem). Transmission lines are also considered in that paper and discussed in Section 4.3. A major message of Beckmann and Wood's paper is the limited performance potential and inordinate complexity of dynamic-NUCA, an innovation that had largely positive attributes in a single-core setting.

Huh et al., ICS'05

In a paper that appeared shortly after Beckmann and Wood's, Huh et al. [22] validate many of the above observations. Huh et al. evaluate a 16-core chip multiprocessor with a large banked NUCA cache in the middle of the chip (see Figure 1.5b). They confirm that an S-NUCA policy leads to long access times on average. If ways are distributed across banks and blocks are allowed to migrate between banks with a D-NUCA policy, there is a minor performance improvement. They consider both 1- and 2-dimensional block movement where 1-D movement prevents a block from moving out of its designated column. The lack of substantial performance improvement from D-NUCA is primarily attributed to the complexity entailed when searching for a block. To avoid having to access numerous banks, Huh et al. implement a distributed set of replicated partial tags. At the top/bottom of every column of banks, partial tags for every block in that column are stored. A look-up into this storage reveals if one or more banks in that column can possibly have the requested block. These additional look-ups of partial tags and banks (nearly 50% more than the S-NUCA case) negate half the benefit afforded by D-NUCA's data proximity. They also result in increased power and bank access rates. The various tag stores will have to be updated with on-chip messages every time a block is replaced/migrated. Huh et al. evaluate L1 prefetching and show that it is an effective technique to hide L2 access times with and without D-NUCA.

The paper also formulates a cache implementation that can operate at multiple points in the shared-private spectrum. *Sharing Degree (SD)* is defined as the number of processors that share a portion of cache. At one end of the spectrum is a cache shared by all processors (an SD of 16 in Figure 1.5b), and at the other end is a cache partitioned into 16 fragments, each fragment serving as a private cache for a core (an SD of 1 in Figure 1.5b). When the cache is completely shared (SD=16), coherence must only be maintained between the cores' private L1s, and directory state is maintained along with each L2 block. When the cache has sharing degree less than 16, coherence must also be maintained among the multiple L2 fragments. A directory is maintained in the middle of the chip and must be looked up after every miss in an L2 fragment. Huh et al. show that different

---

[1] In Section 4.3, we describe the novel Nahalal layout of Guz et al. [6] that reduces latency for shared blocks in a similar D-NUCA architecture.

applications yield optimal performance with different sharing degrees and argue for a dynamic scheme that selects the value of SD on a per-application basis and even on a per-block basis. The logic for such reconfiguration is expected to be non-trivial, especially since per-block reconfiguration may require complex indexing mechanisms.

The rationale for variable sharing degree has its roots in the inherent trade-offs between shared and private caches, explained previously in Table 1.1. As sharing degree is increased, the cache offers better hit rates, but S-NUCA's somewhat random placement of blocks within the shared cache leads to increased cache access times. The work by Huh et al. thus puts forth a reasonable approach to balance the pros and cons of not only shared/private caches, but also S-NUCA/D-NUCA. By employing S-NUCA and a tunable sharing degree, they eliminate complex search, allow for manageable access times, and allow manageable hit rates. The primary overhead is the indexing logic that must be reconfigured and the centralized L2 directory capable of maintaining coherence among 16 L2 fragments. Huh et al. believe that a simple S-NUCA design with a fixed sharing degree of 2 or 4 provides the best combination of performance and low complexity. As we will see later in this chapter, page coloring is an alternative approach to form shared/private regions that eliminates the need for hardware-based L2 coherence, but it relies on software OS support.

Liu et al., HPCA'04

The work of Liu et al. [26] was among the first in the area of shared LLC for multi-cores. Their work did not consider a NUCA cache, but the look-up mechanism introduces non-uniformity in access latency. The LLC is split into multiple banks, all connected to a bus shared by all the cores (similar to Barroso et al.'s Piranha processor [21]). In essence, each bank contains some ways for every set in the LLC. When a core initiates an L2 access, the ways in a subset of banks is first looked up (phase 1). If data is not found here, the ways in the remaining banks are looked up (phase 2). This is similar to the multicast search first introduced by Kim et al. [1]. To allow for the varying needs of each core, Liu et al. allow each core i to look up a certain collection of W_i ways in phase 1. The OS determines this collection of W_i ways for each core by setting up a hardware table. Liu et al. also observe that such an organization (similar to the designs later introduced by Beckmann et al. [2] and Huh et al. [22]) provides a hybrid between a shared and private cache because, (i) similar to a private cache, each core can have its data localized to a few banks, and (ii) similar to a shared cache, hit rates are high as data is not replicated and LLC space can be non-uniformly allocated to cores based on need.

Dybdahl, HPCA'07

Another hybrid shared/private LLC was proposed by Dybdahl and Stenstrom [27]. They assume an LLC that is shared and distributed (similar to Figure 1.6). The ways of a set are distributed across the many slices, so it is a D-NUCA implementation and subject to D-NUCA problems (complex search and migration). Their key innovation is the reservation of some ways in each slice as being private to the local core. The other ways are deemed shared. For any access, the private ways of the core are first looked up, followed by all other ways (similar to the policy of Liu et al. [26]). On every access, the requested block is brought into that core's private ways within the LLC. If a block must be evicted from the private way, it is spilled into one of the many shared ways (not necessarily to a way in that slice). Each core is allocated a certain number of pre-specified ways in the cache (this is heeded when a block from the shared ways must be evicted). This designation of ways as being private and shared results in a replacement policy different than that employed in prior D-NUCA designs. Private ways are also allocated per core based on various run-time statistics. These statistics include the number of hits to the LRU way within the core's private quota (to estimate the increased miss rate if the core was assigned one less way) and the number of hits in shadow tags that track the last evicted block from each private domain (to estimate the increased hit rate if the core was assigned one more way). Such statistics have also been employed by other papers and will be discussed in more detail in Chapter 3.

By introducing a miss-rate-aware allocation of ways to cores, Dybdahl and Stenstrom prevent the shared cache from letting LRU dictate the partition of ways among cores. The way allocations are applied uniformly to all sets and therefore represent a relatively coarse-grain partition.

NuRAPID and CMP-NuRAPID, Chishti et al., MICRO'03 and ISCA'05

While D-NUCA is gradually being acknowledged as "too complex to implement", some early work by Chishti et al. [24] considered an alternative that was higher-performing and potentially less complex in a single-core setting. Their paper in MICRO'03 focused only on a NUCA architecture for a single processing core, while their ISCA'05 paper considers multiple cores.

Two key contributions were made in the first paper: (i) Instead of co-locating tags and data blocks, Chishti et al. propose implementing the entire tag array as a centralized structure near the processing core/cache controller. The cache access begins with a tag look-up and the request is then directly sent to the NUCA bank that has the data. Such a design eliminates the need to search for a block by progressively looking up multiple banks. However, every block movement will require that the centralized cache controller be informed so that tags can be updated. The overhead of tag storage and tag update will increase dramatically in a multi-core setting. (ii) Chishti et al. propose decoupling tag and data block placement. A block is now allowed to reside in any row in any cache bank and the tag storage (organized in a conventional manner) carries a pointer to the data block's exact location. As blocks are accessed, they may move to a bank that is closer to the cache controller, swapping places with any block that may not have been recently touched. Since a swap can now happen between any two blocks, the block movement policy allows the closest banks to accommodate the "globally hottest" (most frequently and recently touched) blocks, and not just the hottest blocks in each set. Note that conventional D-NUCA would restrict each set to only place a small subset of ways close to the CPU, whereas Chishti et al.'s NuRAPID policy (Non-Uniform access with Replacement And Placement usIng Distance associativity) allows all the ways of a hot set to be placed in a nearby bank. Such flexibility can allow NuRAPID to out-perform D-NUCA, especially if applications non-uniformly stress their sets. It can also reduce inter-bank traffic, especially if banks are sufficiently large. The overhead in providing such flexibility is that data blocks need to store reverse pointers that identify their entry in the tag array, so that the corresponding tag can be updated when a swap happens. We may also have to independently track LRU among the rows of a large data bank, although the paper claims that random selection of blocks during a swap is adequate.

Thus, the paper puts forth an interesting organization that decouples tags and data blocks, and allows for more flexible movement of data blocks. Such a decoupled organization was also adopted by more recent proposals to reduce conflicts and improve hit rates ([28, 29], discussed in Chapter 3). It does not suffer from a complex search problem because, similar to Kim et al.'s partial tag storage [1], the entire tags are first examined at the cache controller before sending the request to a unique bank. However, the flexibility entails increased implementation complexity and block look-up in a multi-core setting continues to be problematic.

In a follow-on paper, Chishti et al. extend their scheme to handle multiple cores [23]. Just as in the NuRAPID [24] design, the CMP-NuRAPID [23] design also decouples data and tag arrays. The data array is a shared resource; any core can place its data in any row of the data array; the data array is organized as multiple banks with non-uniform access times. Those are the similarities with the NUCA shared caches described so far in this section. The rest of the design represents an organization with private L2 caches. Each core maintains a private tag array, with entries capable of pointing to any row of the shared data array. Keeping the tag arrays coherent is tricky: Chishti et al. assume that the tag arrays are kept coherent upon misses/movements/replacements by broadcasting changes to all tag arrays. Tag arrays must be sufficiently large to allow substantial coverage of the shared data array and allow for significant sharing of data blocks. For a 4-core processor, the tag arrays combined required twice as many entries as the shared data arrays. This limits the scalability of the approach. The proposed design overcomes three disadvantages of a private L2 organization: (i) If multiple cores read the same data block, data copies are created in each private cache, leading to lower overall capacity. (ii) If multiple cores read/write the same block and have copies in their respective private L2 caches, there are frequent L2 coherence misses, invalidations, and duplicate copies. (iii) L2 cache capacity is statically partitioned among cores.

Chishti et al. address problems (i) and (ii) by maintaining a single copy of a shared block in the shared L2 and having multiple tag arrays point to this block. Problem (iii) is addressed by maintaining large tag arrays that allow a core to map blocks to more than its fair share of L2 space (limited only by the size of the per-core tag array). Data blocks maintain a single reverse pointer to a tag entry – hence, any block movement requires a broadcast so all tag arrays can update their forward pointers. In addition, controlled replication of read-only blocks is allowed. The management of these read-only blocks is similar to their management in a private cache organization: each tag array can point to a copy of the block in a nearby data bank, and a write requires a bus broadcast to invalidate other cached copies and upgrade the block's state from Shared to Modified. Replication of blocks in a shared cache is discussed in more detail in Section 2.1.2.

The CMP-NuRAPID design is therefore an interesting hybrid between private and shared L2 caches. It has much of the performance potential of a shared cache, plus it allows selective replication of read-only blocks (allowing it to out-perform traditional shared NUCA caches). But it also entails much of the complexity associated with private L2 caches, especially that involving tag maintenance and coherence. In that sense, CMP-NuRAPID inherits the primary disadvantage of D-NUCA: if blocks are allowed to move arbitrarily and have no fixed home in the data array, effort has to be expended to either search for this block when an access is desired (D-NUCA), or effort has to be expended so every core has accurate pointers to the block (CMP-NuRAPID).

Innovations for D-NUCA Block Search

We have already mentioned the several problems involved in locating a block in a D-NUCA cache. The original NUCA paper by Kim et al. [1] proposed the use of partial tags at the core to locate the block with modest storage overheads. Extending this solution to a multi-core setting is problematic because every core must now replicate the partial tags and every block movement triggers a broadcast to update these partial tags. Huh et al. [22] adopt a layout and block placement policy that makes search more manageable. Cores constitute the top and bottom rows of a grid network, and a block is restricted to be in its statically assigned column. Requests from cores are routed horizontally to the top or bottom of the column where they consult the partial tags for banks in that column and then route the request to the appropriate set of banks. Chishti et al. [23, 24] assume that the tag array (for most of the L2 cache) is located with each cache controller. The tag look-up precisely identifies the bank that caches the data block and minimizes bank activity. But this presents the same problems as with replicated partial tags: high storage requirements and broadcasts to update tags on every block movement.

Ricci et al. [30] use a prediction based method to narrow down block search without incurring high storage or network overheads. They rely on a Bloom filter [31] at each core to compactly represent the possible location of a block in each bank. While this reduces storage overhead by nearly a factor of 10, compared to a partial tag structure, it introduces a non-trivial number of false positives. Further, Ricci et al. update each Bloom filter only when that core accesses a block, *i.e.*, all Bloom filters are not updated on a block fetch or movement. While this reduces on-chip network traffic, it introduces false negatives because the filters occasionally fail to detect the presence of a block in a bank. Thus, this Bloom filter based mechanism yields a prediction for possible location of a block, and mis-predictions are not uncommon. On an L2 access, the banks indicated by the Bloom filters are first searched; if the block is not located, all other banks must be looked up before signaling a miss.

In spite of these innovations, the search mechanism continues to represent a challenge for D-NUCA. The quest for effective D-NUCA search mechanisms appears to have lost some steam. Not only is the complexity of D-NUCA migration and search difficult to overcome, but there appear to be other solutions on the horizon (page coloring applied to S-NUCA) that combine the best properties of S-NUCA and D-NUCA as well as the best properties of shared and private caches. This approach is discussed shortly in Section 2.1.3.

2.1.2   REPLICATION POLICIES IN SHARED CACHES

In a system with private L2 caches, each L2 cache is allowed to keep a read-only copy of a block, thus allowing low-latency access to the block. In a system with a shared L2 cache, the single L2 can only maintain one copy of each block. This copy may not be in close proximity to cores accessing the block, leading to long access latencies. This is especially true in early S-NUCA designs where block placement is determined by block address bits and the assigned bank can be somewhat random. This problem can be mitigated by allowing the single shared L2 cache to maintain multiple copies of a block, such that each core can find a copy of the block at relatively close proximity. Such replication of cache blocks would then require a mechanism to keep the replicas coherent.

Victim Replication, Zhang and Asanovic, ISCA'05

Zhang and Asanovic propose a simple and clever mechanism to implement block replication without incurring much overhead for coherence among replicas in the L2 cache. As a design platform, they assume a shared L2 cache that is physically distributed on chip, similar to the tiled layout discussed previously in Figure 1.6. An S-NUCA design is adopted, so each block maps to a unique tile/bank (referred to as the *home* bank). Replication is effected as follows: when a block is evicted by a core's private L1, it is placed in the local L2 bank, even though it is not the home bank for that block. This new L2 copy of the block is referred to as a *replica*, and we refer to the corresponding core as the *replicating core* for that block. Luckily, since the block was recently cached in the replicating core's L1, the directory (maintained along with the copy of the block in the home bank) still lists the replicating core in the list of sharers. So, a subsequent write to the block (by any core) will trigger an invalidation message to the replicating core/tile. On receipt of an invalidation, a core must look for that block in its L1 and its local L2 slice and invalidate any copy it finds. This is a very clean way to maintain coherence among L2 replicas, by falling back upon already existing state for L1 coherence and requiring an extra L2 bank look-up on every invalidation. The rest of the coherence protocol remains unchanged.

When a core has a miss in its private L1, it looks up its local slice of L2 (to find a potential replica) before forwarding the request to that block's home bank. Thus, an additional L2 look-up is incurred in this design if a local replica does not exist; conversely, if a local replica is found, there is no additional L2 bank look-up and there is also a saving in on-chip network traffic. If a replica is found in the local L2 slice, it is brought into L1 and removed from the local L2 slice.

The design is referred to as *Victim Replication* because replication is invoked for blocks being evicted out of L1. Not every evicted block is replicated in the local L2 slice. There is of course no replication if the local L2 slice happens to be the block's home bank. There is also no replication if the local L2 slice has no spare room, *i.e.*, the replica can only be created in a location currently occupied by an invalid line, an existing replica, or a home block with no sharers.

Cache indexing is impacted by the Victim Replication scheme. Typically, the most significant bits of the index represent the bank number. When looking for replicas in the local bank, the bank number must be ignored while indexing (since it points to the home bank) and must instead be included as part of the tag to correctly signal a hit on the block. This results in a minor growth in the tag array.

Thus, Victim Replication attempts to combine the favorable properties of shared and private caches. By allowing a large fraction of accesses to be serviced by replicas in the local L2 slice, L2 access time is similar to that of a small private L2. The design continues to enjoy the low miss rate advantages of a shared cache because replicas are only created if there is "spare room". In principle, the design is not dis-similar to having a separate victim cache for each L1. While traditional victim caches are small and fixed, in the proposed design, victim blocks can occupy a large region in the local L2 slice and are careful to not evict other useful L2 blocks. Note that Victim Replication only applies to S-NUCA designs where each core has a designated local L2 slice. It may be employed with a few modifications in D-NUCA designs as long as replicas are not allowed to migrate (to preserve the accuracy of the directory). The value of Victim Replication is somewhat reduced in an S-NUCA design that employs smart page coloring (discussed in Section 2.1.3).

ASR, Beckmann et al., MICRO'06 and CMP-NuRAPID, Chishti et al., ISCA'05

There has been little follow-on work to the notion of replication within a shared cache. This may attest to the fact that the Victim Replication idea leaves little room for improvement in terms of performance and implementation complexity. We briefly discuss two related bodies of work here.

Beckmann et al. [20] propose a dynamic probabilistic system to select the appropriate level of replication. Increased replication leads to lower average access times for L2 hits, but it also lower effective capacity. A block is replicated with a probability N in the local L2 slice (either shared or private) when it is evicted out of L1. The probability N is either increased or decreased periodically based on a cost-benefit estimation. Various structures are introduced to approximately estimate the change in local hit rates and memory accesses assuming a lower or higher probability level N. More details are discussed in Section 2.2. However, these schemes are more applicable to private cache organizations. The Victim Replication policy already has an implicit cost-benefit analysis encoded within it: the replica is created only at the expense of a block deemed dispensable. We could augment the policy to probabilistically allow a replica to replace an "indispensable" block, but there might be little room for improvement (this gap has not been quantified to date).

The second related body of work that allows replication in a shared cache is the previously described work of Chishti et al. [23]. Recall that they employ an unusual design with private per-core tags and a shared data array. Chishti et al. employ replication when an L2 data block is read multiple times by a remote core. When a block is replicated, the design very much resembles a private cache organization. Each core's private tag array points to a nearby copy of the data block. Since the tag array stores the block in Shared state, a write to that block is preceded by an invalidation operation on the bus to get rid of other shared copies. The reason this discussion finds itself in the shared-cache section of this book is because they allow multiple tag arrays to point to a single shared block in the data array (to not only simplify producer-consumer sharing but also boost capacity by avoiding replication at times).

SP-NUCA and ESP-NUCA, Merino et al., HPCA 2010

The SP-NUCA architecture of Merino et al. [32, 33] introduces separate indexing functions into a shared cache for each core. This allows them to design a hybrid *Shared-Private NUCA (SP-NUCA)* architecture. In their implementation, the last level cache is organized as a shared S-NUCA. A block is first assumed to be a private block and only the local bank is looked up (the bank number bits of the address are not used for indexing and are stored as part of the tag). If the block is not found in the local bank, it is assumed to be shared and regular S-NUCA indexing is employed (the bank number bits of the address are used to send the request to the appropriate bank). If the block remains unfound, the private banks of every other core are looked up before sending the request to memory. During its L2 lifetime, a block is initially classified as "private" and then may transition to a different bank in "shared" state if it is touched by another core. While SP-NUCA is efficient at localizing private data, it requires multiple bank look-ups to locate shared data. If the workload is entirely multi-programmed, SP-NUCA degenerates to behavior similar to a private cache organization (higher miss rates). In addition to the limited space for private data, SP-NUCA also suffers from longer latencies for shared blocks. Merino et al. [33] introduce the ESP-NUCA architecture to alleviate these concerns. Some shared blocks are allowed to be replicated. In addition, ways of a bank can be dynamically allocated across private, shared, and replicas. Techniques similar to dynamic set sampling (discussed later in Section 3.1.1) are used to identify the best partition.

2.1.3   OS-BASED PAGE PLACEMENT

A shared L2 cache with an S-NUCA policy has several desireable features: it minimizes L2 cache miss rates and allows for simple block look-up. Most early work in the area pointed to a major short-coming in such a design: blocks are statically placed in banks based on the block's physical address; assuming that block physical addresses are somewhat random, blocks tend to get scattered uniformly across the many banks. As a result, on average, a request is serviced by a bank that is half-way across the chip. This motivated the several bodies of work on D-NUCA that allow a block to reside in any one bank and attempt to bring relevant blocks to closer banks. This also motivated several studies on private cache organizations because a small private cache has a lower hit time than the average hit time of a large shared S-NUCA cache.

However, unlike the assumption in early work, block physical addresses need not be random. The use of OS-based page coloring [34] can control block physical addresses and hence their placement in banks. These principles have been known for several years and even been employed to improve locality in cc-NUMA multiprocessor systems in the 1990s [35, 36, 37, 38, 39, 40]. This idea remained untapped within the NUCA space until a paper by Cho and Jin in MICRO 2006 [41]. In the last few years, it is becoming increasingly clear that this approach perhaps combines the best of S-NUCA and D-NUCA, the best of shared and private caches, and balances performance and implementation complexity.

Cho and Jin, MICRO'06

We begin the discussion by examining how data blocks get mapped to banks in an S-NUCA cache. Figure 2.2 shows a 32-bit physical address and the bits used to index into a 16 MB 8-way L2 cache partitioned into 16 banks. This example can be easily extended to depict the larger physical addresses that are typical in modern systems. If we assume a 64-byte cache line size, the last (least significant) six bits (offset) are used to select individual bytes within a cache line. The cache has 32K sets, each set containing 8 ways. Fifteen bits are required for the set index. In an S-NUCA cache,

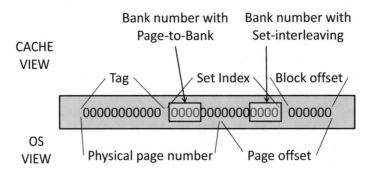

Figure 2.2: The interaction of physical address bits with cache placement and page mapping.

sets are distributed across banks, but an entire set (all 8 ways) resides in a single bank. Hence, a subset of the set index is used to identify the bank that contains the block. Early studies assumed that the least significant bits of the set index are used as the bank number. We will refer to this as *set-interleaved mapping*. This causes consecutive sets to be placed in different banks and tends to uniformly distribute program working sets across all banks. This helps distribute load across banks and reduces contention and capacity pressure at each bank. Unfortunately, this means that requests from a core are just as likely to be serviced by the most distant bank as they are to be serviced by the nearest bank.

Alternatively, if the most significant bits of the set index are used as the bank number, several consecutive sets will be placed in the same bank. We will refer to this as *page-to-bank mapping* because an entire OS page can potentially be placed in a single bank. Because programs exhibit spatial locality, a series of requests to the same page will be serviced by the same bank, leading to high contention and uneven capacity pressures.

Now consider the operating system's view of the block's physical address (also shown in Figure 2.2). If we assume that the OS page size is 16 KB, the least significant 14 bits of the address represent the page offset, and the remaining bits represent the physical page number. When the OS assigns a program's virtual page to a physical page, it ends up determining a subset of the set index bits for all cache lines within that page. If we employ *set-interleaved mapping*, the page number bits (for the example in Figure 2.2) do not intersect with the bank number bits; as a result, the assignment of a virtual page to a physical page does not determine a unique bank that the page gets mapped to. The page gets distributed across all banks. If we employ *page-to-bank mapping*, the bank number bits are a subset of the page number bits. Thus, by assigning a virtual page to a physical page, the OS also determines the cache bank that the entire page gets mapped to, hence the term *page-to-bank mapping*. Such a mapping presents the option to leverage the OS to optimize data placement in a shared cache, instead of the hardware-intensive and complex D-NUCA policies. Adopting standard terminology, the bank number bits, that represent a subset of the page number bits, can be referred to as the *page color*. If there are 16 banks (and hence, 16 colors), the OS organizes its free pages into 16 lists based on color. When a new page is requested, the OS first determines the optimal bank for that page and then selects a free page from the appropriate color list. By coloring the page appropriately, the OS ensures that the entire page is cached in the optimal bank.

To recap, we are employing a large shared cache with an S-NUCA policy and an indexing mechanism that maps an entire OS page into one bank. It is expected that future shared caches will be distributed on chip, with one or more banks co-located with each core (Figure 1.6). Such a layout, combined with the appropriate page coloring policy can also mimic a private cache organization for many workloads. If we assume that threads running on different cores access mostly private data, *i.e.*, the sets of pages touched by the cores are largely disjoint, then the OS can color pages so they are placed in the local banks of the accessing core. Each L2 bank can therefore be made to serve as a private cache for its corresponding core. In fact, such a design can out-perform an organization with traditional private L2 caches (Figure 1.3) because on an L2 miss, the request can be directly sent to the next level of the hierarchy without having to check the contents of other private L2 caches. This design does deviate from a private cache organization when handling a shared page. A shared page will have a unique instance in the shared L2 cache and will reside in exactly one bank. Therefore, unlike the private cache set-up, the block cannot be placed in the local bank of each core that accesses the block. This can be detrimental for performance if the blocks are primarily read-only.

As mentioned above, an obvious OS page coloring policy, that has also been used for cc-NUMA multiprocessors [36, 38, 39, 40], is *first-touch* color assignment. In this policy, a page is assigned a color that places it in the cache bank nearest to the core that first touches that page. It is based on the premise that most subsequent accesses to that page will also be from the same core. This is certainly a valid premise for a workload that primarily consists of single-thread programs. It is also a valid premise for multi-threaded applications where a majority of the working set happens to be private to each thread.

Let us first consider a workload that consists of single-thread programs. If we assume that every block is only touched by one core, then the first-touch color assignment allows every L2 access to be serviced by the local bank, leading to low L2 access times. But, correspondingly, each core only enjoys a fraction of the total shared L2 capacity since all of its data is steered towards a single bank. This behavior strongly resembles that of a private cache organization. In order to out-perform the private cache organization, the page coloring policy must be augmented so each core can have ownership of a larger fraction of the shared L2 if required, without incurring long L2 latencies.

Cho and Jin [41] propose *page-spreading* policies that augment a baseline first-touch page coloring policy. When a cache bank experiences high "pressure", subsequent page requests from that core are assigned colors that map the pages to neighboring banks. This allows a core to have more than its fair share of L2 cache space, which is the fundamental reason behind a shared cache's superior hit rates. Since the policies are attempting to localize a core's working set in the adjacent and neighboring banks, L2 access times will continue to be relatively short. Cho and Jin propose the use of Bloom filter based counters to track the number of unique pages accessed recently by a core; the counter value serves as the metric for cache pressure. Awasthi et al. [42] formalize and evaluate Cho and Jin's page-spreading approach. In essence, every page request is mapped to a bank that minimizes a mathematical function that estimates cache pressure and the distance between the bank and requesting core. As a result, cache hit rates approach that of a baseline set-interleaved shared cache, and most L2 requests are serviced by the adjacent or nearby banks. Such a design is expected to out-perform a private cache organization on practically all workloads comprised of single-thread programs (a few caveats mentioned subsequently).

In a multi-threaded workload, each page should ideally be placed in a bank that represents the center-of-gravity of all requests to that page. The first-touch policy (with or without page-spreading) may end up placing a shared page in a sub-optimal bank. This can be remedied by migrating the page over time, and this approach is discussed shortly. Without such dynamic migration, the behavior for shared pages is likely to be very similar to that in the baseline set-interleaved shared L2 cache. The first-touch policies will continue to provide low access times for the many private pages in such workloads.

In summary, the use of an S-NUCA shared cache combined with OS page coloring policies is perhaps a compelling design point when considering performance and implementation complexity. It provides most of the benefit of D-NUCA policies without much complexity overhead; it therefore represents a favorable design point between the original set-interleaved S-NUCA and D-NUCA. In addition, the proposed OS-based approach on an S-NUCA shared cache provides the biggest advantage of private caches over shared caches: quick access to private data blocks. The page-spreading policies also allow the cache to meet the high hit-rate potential of baseline set-interleaved shared caches. Note, however, that a private cache organization can out-perform the proposed OS-based approach when handling some multi-threaded applications: in the former, a shared block may frequently be found in the local private cache, while in the latter, the shared block will have a single (possibly distant) residence.

In follow-on work, Jin and Cho [43] employ compiler hints to guide the OS in selecting a near-optimal page color. At compile time, the program is executed with a test input and access counts per page per core are collected. Pages are clustered into groups, and the access pattern for the group is matched against a handful of well-known access patterns. Based on the access pattern, a hint is included in the binary to guide the OS to map each page to a tile that is expected to be the dominant accessor of that page. It is not clear if the compiler hints provide significantly more information than that provided by first touch. Some pages are marked as being shared equally by all cores, and such pages are scattered across all tiles by modifying the cache indexing function for that page. Jin and Cho also make an initial case for replication of shared pages, although coherence issues for these replicated pages are not considered.

A paper by Marty and Hill [44] mentions that coherence overheads in a multi-core chip can be high if remote directories must be contacted on every L1 cache miss. This is especially wasteful when cores are allocated across virtual machines (VMs) and valid data is likely to be found in a nearby tile. Marty and Hill solve the problem with a two-level coherence protocol where a first-level protocol stores a directory within the cores used by a VM. A miss in this directory invokes the more expensive second-level protocol. This is an alternative approach to solving the locality problem in an S-NUCA cache. However, the use of page coloring, as suggested by Cho and Jin [41], would largely solve the locality problem with little hardware support and not require a two-level hierarchical coherence protocol. In other words, if first-touch page coloring was employed, most pages would be cached in tiles that belong to the VM; the overheads of accessing the LLC bank and enforcing directory-based coherence would therefore not be very high.

Page coloring was also employed by Madan et al. [9] to reduce communication overheads in a 3D reconfigurable S-NUCA cache. This work is described in more detail in Section 5.4.

Awasthi et al. and Chaudhuri, HPCA'09

Cho and Jin's first-touch page coloring policy combined with page-spreading is very effective at balancing the latency and capacity needs of private pages. It is not very effective at finding an optimal home bank for shared pages. It is also not effective if threads migrate and leave their working sets behind in their original banks. Depending on how a program moves through different phases, it is also possible that a thread may cache relatively inactive pages in its local bank and "spread" more active pages to neighboring banks, resulting in a sub-optimal mapping of data to banks. Other dynamic conditions (such as programs finishing or starting) may also cause fluctuations in cache bank usage, rendering previous page mapping decisions as sub-optimal. In order to deal with the above weaknesses, we may need dynamic policies that track cache usage and affect page migrations if the "optimal" home bank of a page is different than the one that was estimated on first-touch. This migratory approach would start to resemble the already discussed D-NUCA policies. In order to make this happen, a page must be copied in DRAM from its current physical address to a new physical address that has the appropriate optimal color. In addition, all traces of the old physical page must be wiped out of the old cache bank and out of the TLBs of all cores.

Awasthi et al. [42] show that migrations are helpful as first-touch decisions tend to become sub-optimal over time, especially for multi-threaded applications with many shared pages. In order to avoid the expensive copy of pages in DRAM, they propose a mechanism that introduces another level of indirection on chip. Consider a page that is initially placed in physical address A and is mapped to some bank X. In order to move the page to a different bank Y, the on-chip TLBs are updated to refer to this page by a new physical address A'. But the page continues to reside in the same physical address A in DRAM; it just has a new name on chip (L1 and L2). If a block is not found in L2 and has to go off-chip to memory, a request for the original physical address A is issued to DRAM. When a page is renamed from A to A' on chip, the shadow address region is used for A' to ensure that the new name is not already being used by another page. The shadow address region represents addressable portions of memory that do not actually exist in DRAM; for example, in a 32-bit system with 2 GB DRAM, the address range 2-4 GB is the shadow address space. A' is selected such that it has the appropriate new page color, and the original page color is also encoded into A'. This makes it easy to re-construct A from A', an operation required when there is an L2 miss and the request must be issued to DRAM. The primary overhead of this on-chip renaming mechanism is a relatively large on-chip structure, called the *Translation Table*. It keeps track of every page migration ($A \rightarrow A'$) and is analogous to a large on-chip second-level TLB. Note that every page migration continues to require a flush of the original page's contents from cache and from TLBs.

In addition to this low-cost migration scheme, Awasthi et al. introduce OS policies that periodically examine hardware counters and effect migrations. Based on cache usage and hit rates, banks (page colors) are classified as *Acceptors* (need more cache space) and *Donors* (can spare some cache space to other programs). A program that needs more cache space is allowed to map its new pages to a donor color; an overly pressured acceptor color is allowed to move some of its pages to a donor color. Donor colors are chosen by computing a function that considers cache usage and distance between core and banks.

In concurrent work at the same conference, Chaudhuri makes a similar argument [45] with the PageNUCA design. He too shows that page migrations within the L2 are very beneficial for both multi-threaded and multi-programmed workloads. To migrate a page within L2, the OS-assigned address (P) is converted into a different address (P') with the appropriate page color. Since shadow addresses are not used, it is possible that another virtual page may be mapped to P'; this page must be swapped into the address P. Since the L1 cache and main memory continue to use the old mappings, tables are required at cores (dL1Map and iL1Map) and at the L2 (forward and inverse L2Maps) to perform translations back and forth between P and P'. Two techniques are proposed to identify pages that need migration: (i) a hardware-based technique that dynamically tracks access patterns with a page access counter table (PACT), and (ii) a manual code instrumentation process that designates page affinity to cores at the start of the program. When a page is identified for migration, the L2 is locked up while blocks are explicitly copied between the source and destination banks, and various map tables are updated (this is unlike the technique of Awasthi et al. where the original page blocks are invalidated and the new bank is populated on-demand as blocks in the new page are touched). Overall, a fair bit of complexity is introduced to determine the pages that require migration and to keep the various tables updated on a migration. Chaudhuri makes the claim that page granularity is appropriate for migration (compared to D-NUCA's block granularity) because (i) a page migration essentially performs prefetched migration for several blocks in that page, (ii) a bulk transfer is more efficient because it can be efficiently pipelined across a scalable interconnect, and (iii) overhead in all book-keeping tables is reduced.

In the real world, it is highly likely that such migratory schemes will be required to adapt to continuously changing workloads and operating conditions. These migratory schemes are an attempt to take the previously proposed Cho and Jin approach (S-NUCA with first-touch page coloring) even closer to D-NUCA. Migration does involve some overheads, and it is not clear if these will ever be tolerable. The papers by Awasthi et al. and Chaudhuri attempt to alleviate one of these overheads (DRAM copy) but end up introducing different non-trivial overheads (Translation Table, L1/L2Maps, etc.). Hence, the quest for an elegant migratory scheme continues.

Reactive NUCA (R-NUCA), Hardavellas et al., ISCA'09

In recent work, Hardavellas et al. [3] put forth a novel NUCA architecture that relies on OS management of pages in a large shared L2 cache and does not require complex search mechanisms. While this resembles the same objective of Cho and Jin, Hardavellas et al. adopt a different approach that is centered around classifying the nature of a page and handling each class of pages differently. Each core is also allowed its own indexing functions, enabling each core to have a different view of the shared L2 cache – this allows a private page to migrate between banks without requiring page copy in DRAM or complex hardware structures. In addition to efficiently handling both shared and private pages, it facilitates replication at various granularities. While there is no explicit focus on migration and cache capacity allocation among competing threads, their approach can be extended to handle those problems. It therefore represents one of the best design points today. The above properties will be clarified in the subsequent discussion.

The first observation the authors make is that pages can typically be classified into three categories.

(i) *Shared instruction pages:* These pages are typically accessed by all threads in a multi-threaded application in read-only mode. Such pages can be replicated in each L2 cache bank without requiring a coherence mechanism within the shared L2. To prevent high cache pressure from replication, a page is replicated once in a cluster of I banks, where I can be potentially determined at run-time (I is assumed to be 4 in the paper [3]).

(ii) *Private data pages:* These pages are private to each thread and must only be placed in the local bank. To handle capacity allocation of space among heterogeneous threads, the policy can be modified to allow private pages to be placed in one of P local/nearby banks (the exact indexing mechanism will be described shortly), where P can be determined at run-time for each thread (P is set to 1 in the paper for all threads).

(iii) *Shared data pages:* For server workloads, data pages are typically shared by all threads and exhibit reads and writes. Each block must therefore be instantiated in a single bank (to avoid need for coherence), and pages can be equally distributed across S banks (to minimize miss rates and hotspots and because it is difficult to estimate an optimal bank for each page). S is set to 16 (the number of cores) in the paper.

Before we consider how blocks are placed in the cache, let us first describe the process for classifying pages into the three categories above. The process is entirely OS-based and requires book-keeping in the page tables and TLBs. On first-touch, a page is marked as either being an instruction page or a private data page, depending on whether it originated from the L1 instruction or data cache. If a core has a data TLB miss and the OS recognizes that the page is marked as "data private" elsewhere, it re-classifies the page as being "data shared" and flushes the page out of the other core's TLB and L2 cache bank.

Next, consider data placement in banks. As an example, assume that banks are organized as clusters of size n. Assume that each cluster is disjoint. Once a page is assigned to a cluster, it is placed in one of the n banks depending on the page address bits. All cores within that cluster recognize a uniform numbering scheme for the banks in that cluster, so given an address, any of the cores can easily figure out the bank that stores the page. To maximize proximity of data to the requesting core and to balance load across clusters, Hardavellas et al. use a clever rotational interleaving mechanism. Each core is assumed to be part of a cluster with the core at its center (thus ensuring that the core's data is placed in its immediate vicinity). Clearly, clusters are now overlapping (unless the cluster size is 1). Each bank is given a number between 0 and $n - 1$ (referred to as the rotational ID), such that each cluster ends up having banks with unique rotational IDs (an example is shown for $n = 4$ in Figure 2.3). Since a bank now belongs to four different clusters, its load is roughly the average load

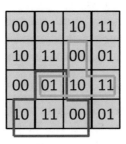

Figure 2.3: An R-NUCA cache with a rotational ID numbering system for $n = 4$ and two example clusters (in red and blue) [3].

in those four clusters (thus leading to better load distribution across banks). Since each bank has a fixed rotational id, all cores again recognize a uniform numbering scheme for the banks, and page requests can be easily directed to the appropriate bank.

As mentioned above, logical clusters of three different sizes are adopted – shared data pages assume that banks are organized as clusters of size 16, instruction pages assume clusters of size 4, and private data pages assume clusters of size 1. So if a core issues a request for a private data page, it knows that the page is placed in the local cluster; since this is a cluster of size 1, the request is sent to the local bank. If a core issues a request for an instruction page, it knows that the page is placed in one of the banks in its local cluster of size 4; two bits of the page address designate this bank (this is the bank's rotational ID) and the request is sent there. Note that other cores that have this bank in their cluster will also access this instruction page in the same way (all these cores refer to the bank by the same rotational ID); so this page is essentially shared by 4 cores. This page may also be present in 3 other banks (in a 16-core processor), thus leading to controlled replication. The cluster size designates the degree of replication (and sharing) for instruction pages. If a core issues a request for a shared data page, it knows that the page is in one of the banks that form a cluster of size 16; four bits of the page address are used to identify this bank and send the request there. The same process is used by all the cores that have the bank in their cluster. Note that each core is aware of the class that a page belongs to (the page class is stored in the TLB) and can easily index into the L2 cache banks in the appropriate way. When a page is re-classified from being "data private" to "data shared", the page will index differently into the cache because of the change in cluster size. Hence, the old contents of the page must be flushed from a single cache bank and reinstated in the new bank (the DRAM physical address remains the same, so no DRAM copy is required).

The R-NUCA architecture has simple indexing, allows controlled replication, and appropriately handles private and shared pages. The additional burden placed on the OS is the book-keeping in the page tables and TLB to track the page classification; no OS-based page coloring is required because it is the page classification and the corresponding cluster size that determines where the page is placed. No auxiliary hardware structures are introduced because there is no explicit migration (apart from the compulsory movement of a page from being "data private" to "data shared"). Once a page is designated as "data shared", migration is deemed as unworthy because shared pages are observed to be roughly equally shared by all (16) cores. While the authors do not evaluate heterogeneous cache capacity allocation across cores, this may be possible by allowing each core to have varying values of P, instead of the default size of 1. However, every time P is changed, all private pages will have to be flushed and migrated to their new locations. Since DRAM physical addresses are never changed, again, no DRAM copies are required. In spite of the overheads to dynamically estimate P and implement the migration, the design does not suffer from the other migration overheads (various tables for translation) present in the work of Awasthi et al. [42] and Chaudhuri [45].

Lin et al., HPCA'08 and SC'09

In a recent study, Lin et al. [46] evaluate the impact of page coloring on shared cache management in a real system with dual-core Xeons. This study serves to validate prior simulation-based studies and reveals a few new observations as well. Since the Xeons do not offer non-uniform cache access, page coloring is not being used to improve data proximity, but to simply enforce shared cache partitioning between two competing cores/programs. A static policy is evaluated where working set sizes are known beforehand, and a dynamic policy is also considered where run-time measurements every epoch guide incremental allocations of page colors to cores. Performance improvements average in the neighborhood of 10%. The dynamic policy requires page migrations in DRAM when a page is re-assigned to a different color. Lin et al. note that this can be a problem and implement a lazy page migration mechanism that limits the performance degradation from page migration to only 2% on average. These overheads may be higher if there are frequent migrations or if there are more cores and greater demands on the memory system. They also note that a program's performance sometimes improves when part of its cache allocation is given to the co-scheduled program. This is because the overall lower L2 miss rate allows the first program to benefit from shorter memory queuing delays.

Lin et al. [47] follow up their HPCA'08 work with a solution that takes advantage of small hardware structures and OS-controlled policies. A large shared UCA L2 cache is assumed, and the cache is partitioned into C colors. A physical page maps entirely to one of these colors. To dynamically control capacity allocation across threads, a page's color can be changed on the fly; the physical address of the page is not changed, just the index bits used to look up the L2 cache. The L2 is preceded by a *Region Mapping Table* that stores the new cache color for each page. Clearly, this can result in a large table; hence, colors are re-assigned at the coarser granularity of a region (about 16 MB in [47]). The table need not be looked up on every access because the new color can also be stored in the TLB. Thus, the *Region Mapping Table* is very similar in spirit to the *Translation Table* of Awasthi et al. [42], but it has manageable complexity because data is re-mapped at a much coarser granularity. The overall approach of using another level of indirection and combining hardware help with software policies is also similar to that of Awasthi et al. [42] and Chaudhuri [45]. Again, the movement of pages does not require copies in DRAM, but it does require cache flushes and TLB shootdowns. The primary advancement in this work is the use of coarse granularity re-mapping of data to cache colors to achieve a solution with good speedup and very low complexity. The re-mapping facilities must be combined with a good hardware profiling unit and software policies to compute optimal assignment of regions to colors. The profiling unit employed by Lin et al. uses shadow tags to estimate miss rates as a function of associativity (shadow tags are discussed in more detail in Section 3.1.2).

2.2 DATA MANAGEMENT FOR A COLLECTION OF PRIVATE CACHES

Having discussed innovations for a shared LLC, we now turn our attention to organizations where the LLC (L2 or L3) is composed of banks that are logically and physically private to each core. Recall from the discussion in Section 1.1 that private caches help keep the most popular data for a core at the nearest cache bank, but they usually offer poor hit rates because of data replication and the static allocation of cache space per core. A miss in a private LLC also typically requires a look-up of other on-chip private caches before the request can be sent off-chip. Most innovations for a private LLC attempt to improve its negative feature (the low hit rates) while preserving its positive attribute (low average latency on a hit). However, one of the negative features (the complex structures required for coherence among private caches) continues to persist in most proposed innovations for a private LLC.

Cooperative Caching, Chang and Sohi, ISCA'06 and Speight et al., ISCA'05

A private cache organization tends to have lower hit rates because each core is statically allocated a fixed size private cache and because of replication of data blocks. Chang and Sohi [48] attempt to relax the first constraint by allowing a cache bank to house a block evicted by another cache bank. The second constraint is alleviated by modifying the replacement policy to favor evicting replicated blocks. Their baseline implementation assumes a directory-based protocol for coherence among private L2s. On an L2 cache miss, a central directory is contacted before forwarding the request off-chip. This directory essentially replicates the L1 and L2 tags for all on-chip cache structures (similar to the description in Section 1.1) and can identify the existence of a dirty copy of the requested block on chip. Chang and Sohi propose three modifications for an on-chip private LLC.

(i) Cache-to-cache sharing: The first modification is well-known but typically not adopted in traditional multi-chip multiprocessors. On a cache miss, the coherence protocol can allow a sibling cache to respond with data if that is faster than servicing the request from the next level of the hierarchy. This was often not true in traditional multi-processors where off-chip and network access was required for the sibling cache and the next level. But in modern multi-cores, access to a sibling cache on the same chip via on-chip networks is much faster than off-chip access to the next level of the hierarchy. Hence, Chang and Sohi advocate the use of cache-to-cache sharing, with the directory being responsible for identifying the sibling cache that can respond and forwarding the request to it.

(ii) Evicting replicated blocks: The second modification is to the private cache replacement policy. When a block is being evicted, the replacement policy first attempts to evict a block that may have replicas elsewhere on chip, thus attempting to reduce the degree of block replication in the private caches. This is not done for every eviction but probabilistically for $N\%$ of all evictions, where N can be determined per workload. In order to effect this change, each cache must have information about whether its blocks are *replicas* or *singlets*. It is the directory's responsibility to gather this information and propagate it to caches every time the status of a block changes. Clearly, this entails non-trivial on-chip traffic, especially since caches can no longer silently evict blocks.

(iii) Spilling of evicted blocks: If a singlet block must be evicted from a private cache, further attempts are made to retain it on-chip. With a given probability M, the block is handed over to a sibling private cache. The sibling cache accommodates the block (hopefully evicting a replica in the process) and marks it as LRU. This buys the block some time to hopefully be accessed by some core and prolong its on-chip lifetime. Bits are maintained along with the block to limit the number of times such a block can be spilled to a sibling (Chang and Sohi employ one-chance forwarding in their evaluation). As always, the directory must be informed of any block movement.

The three modifications are together referred to as *Cooperative Caching*. It can yield improvements over unoptimized shared and private baselines of the time, and this improvement is often a strong function of N and M. Unfortunately, as discussed above, the accompanying overhead for the directory is non-trivial.

The Cooperative Caching framework of Chang and Sohi was preceded by work by Speight et al. [49] that also proposed a form of co-operation among private on-chip caches. It included spilling and cache-to-cache sharing aspects. Speight et al. assume a baseline processor with private L2 caches and a shared off-chip L3 that serves as a victim cache. This platform has various subtleties in the coherence protocol that we will not cover here. Relevant to this discussion is the fact that coherence among L2s and L3 is maintained with a broadcast-based snooping protocol. Since access to other private on-chip L2s is faster than access to either L3 or memory, cache-to-cache sharing is encouraged.

When a block (either clean or dirty) is evicted from a private L2, a writeback is propagated on the shared bus so that the block can eventually be written into the off-chip victim L3. Another L2 is allowed to host this block so that the off-chip access can be avoided. Speight et al. refer to this process as *snarfing* by the other (host) L2s (often referred to as *spilling* in this book). Snarfing is allowed if the host L2 has room to spare, *i.e.*, if the new block only replaces lines that are either invalid or shared. In addition, this is allowed only if the block has previously exhibited potential for reuse. This is estimated by maintaining a table per L2 cache that tracks all recent lines evicted by L2s and if they have been reused again. Both of these events are observed by snooping on the bus. When such a line is evicted again, a bit is set during broadcast that allows other L2 caches to snarf this block. If multiple L2 caches volunteer to serve as hosts, only one is designated as the host. The coherence protocol has a central entity on the bus that collects snoop responses and decides which cache gets to host the evicted line.

The schemes of Speight et al. [49] and Chang and Sohi [48] are similar in spirit. Implementation differences are primarily borne out of the use of snooping in the former case and a directory in the latter. In order to trigger spilling, Chang and Sohi make an effort to identify singlets (even using this information in the replacement policy), while Speight et al. use tables to predict reuse potential.

ASR, Beckmann et al., MICRO'06

The Cooperative Caching mechanism of Chang and Sohi uses run-time parameters N and M to dictate the level of replica eviction and singlet spilling. But they do not consider the run-time estimation of optimal N or M. Beckmann et al. [20] introduce run-time schemes to select the appropriate replication level and these schemes can be applied to any of the already proposed replication mechanisms (Victim Replication, CMP-NuRAPID, Cooperative Caching). In many caching innovations, we must answer questions such as these: when a block is evicted from L1, we must choose to either keep it in the local L2 bank or not (victim replication); or alternatively, when evicting a block from L2, we must choose to either evict an LRU singlet or a non-LRU replica (cooperative caching). Beckmann et al. introduce *Selective Probabilistic Replication (SPR)* that favors replica creation if a pseudo-random number generator is less than a value R. They augment this simple probabilistic scheme with a mechanism to determine the optimal value of R, referred to as *Adaptive Selective Replication (ASR)*. The key contribution of the paper is the design of structures that help estimate the local hit rates and memory accesses if R were increased or decreased. Regardless of whether we consider increasing or decreasing R, there is an associated cost and benefit, and the following four mechanisms aid this estimation:

- **Benefit of increased replication:** A shadow tag structure keeps track of the contents of the cache assuming a higher replication level, thus estimating the additional replica hits that this would lead to.

- **Cost of increased replication:** Hits to LRU blocks in cache are tracked to determine the new misses that might be created by increasing the replication level.

- **Benefit of decreased replication:** Another structure keeps track of recently evicted blocks to estimate if high replication is evicting potentially useful blocks.

- **Cost of decreased replication:** The cache keeps track of hits to replicas that may not have been created at a lower replication level, thus estimating the cost of moving to a lower replication level.

The above estimates help determine if R must be increased or decreased to find the optimal balance between higher local hits and higher effective capacity.

Distributed and Elastic Cooperative Caching, Herrero et al., PACT'08 and ISCA'10

An important drawback of the Cooperative Caching scheme of Chang and Sohi [48] is the need for a complicated centralized directory that may be accessed multiple times on every LLC miss. This centralized directory replicates the tags of every on-chip L1 and L2; it is updated on most block movements; it is responsible for identifying singlets/replicates and identifying opportunities for cache cooperation. When looking for a block, a large number of tags must be searched because the block may reside in any of the ways of the many L1 and L2 caches.

The Distributed Cooperative Caching (DCC) scheme of Herrero et al. [50] attempts to alleviate some of these implementation complexities. First, they propose the use of a conventional set-associative global tag array to keep track of all on-chip blocks. This tag array can have low associativity and a large number of sets and has little resemblance to the replicated tags maintained in Chang and Sohi's Cooperative Caching. This global tag array is expected to track the sharing status of every on-chip block (one bit per private LLC for each tag). Because of the low associativity of this tag array, it is possible that some block's tag may have to be evicted out of this tag array even though it may still exist in some on-chip cache – to prevent such an inconsistency, the block is evicted from all on-chip caches. The hope is that this event can be made relatively infrequent by having many sets in the tag array. This eviction policy is the result of enforcing LRU on the global tag array shared by all cores and can therefore yield better hit rates than the locally made eviction decisions in Cooperative Caching. The new organization is much more energy-efficient because a limited number of tags must be looked up on each access.

Herrero et al. also propose that this tag array be banked and distributed on chip with an interleaving that places successive sets in neighboring banks. This distributes load on the network and removes the centralized bottleneck found in Cooperative Caching.

Note that the proposed solution does not alleviate all of the implementation inefficiencies of Cooperative Caching. Several messages must be exchanged on every block eviction (even clean block evictions) to update the tag arrays and to modify singlet/replicate status.

Herrero et al. augment their design with a mechanism to allocate a cache bank across private and shared blocks [51]. Each cache bank tracks the number of hits in the LRU block among private ways and the hits in the LRU block among shared ways. At periodic intervals, each cache bank informs other cores about the number of ways that have been allocated to shared ways. When blocks are evicted, if keeping the block around is expected to have high utility, it is spilled to one of the globally shared ways in round-robin order.

Dynamic Spill-Receive, Qureshi, HPCA'09

In recent work, Qureshi puts forth an elegant design that builds on the Cooperative Caching framework. Apart from the ASR policy [20], most Cooperative Caching policies, proposed so far, [48, 49, 50] do not estimate the cost/benefit of spilling an evicted line to a neighboring cache. More notably, the neighboring receiver cache does not even have an option and must accept the block that is forced upon it. The ASR policy of Beckmann et al. [20] tries to estimate this cost/benefit and set the probability thresholds to encourage or discourage spilling. Qureshi proposes a solution that simplifies the cost/benefit estimation and that is cognizant of a cache's ability to receive blocks evicted by others.

In Qureshi's *Dynamic Spill-Receive (DSR)* design [52], each private cache is designated as either a *Spiller* or *Receiver* (but not both). The probability thresholds are eliminated. The only cooperation allowed is the spilling of blocks from Spiller caches to Receiver caches. The Spiller/Receiver designation is done in a manner that maximizes on-chip hit rates. Each private cache independently designates itself as Spiller or Receiver with an estimation based on a technique called Set-Dueling [53] (explained further in Section 3.1.1). Each cache always allows spilling for a few of its sets (for example, 32 out of 1024 sets) and always allows receiving for a few of its sets; these groups are referred to as the "always-spill" and "always-receive" groups. A single counter per cache keeps track of which group of sets yields more misses and dynamically selects the better policy (spill or receive) for all other sets in the cache. Note that this counter also tracks misses in these sets in all other caches (easily computable by paying attention to the bus in a snooping-based protocol). As a result, a cache's Spiller/Receiver status is also a function of the other applications that it is co-scheduled with. Each private cache uses different dueling sets to decide on its Spiller/Receiver designation.

The above process is very low overhead, only requiring a single counter per cache. The one-chance forwarding scheme of Chang and Sohi requires a bit per block to denote if the one chance has been used or not. That bit is no longer required as a Receiver cache is not allowed to spill, and a block will almost always be spilled only once. A Receiver cache may become a Spiller cache over time, allowing a block to be spilled multiple times. Statistics collected on the dueling sets can estimate if an application is suffering because of its Receiver status and enforce QoS by switching to Spiller status if necessary.

CloudCache and StimulusCache, Lee et al., HPCA'11 and HPCA'10

Lee et al. [54] assume a processor where the LLC is composed of private caches. They propose the CloudCache design where the boundary of each private cache is not fixed at design time. A core could potentially use ways from a number of cache banks to form its private cache. Depending on the distance of these ways from the core, a priority chain is formed, also doubling up as the priority list used for replacement decisions. As blocks are accessed and garner a higher priority, blocks are shuffled among banks to reflect the priority order. Thus, the ways allocated to a core, potentially spread across multiple banks, essentially form a D-NUCA organization. Block search is done by broadcasting a request to all ways that are allocated to a core. Broadcasts are disabled if another core is accessing the same block via the directory. Each core maintains tags for an additional 32 ways for a few sample sets. This allows the core to determine the utility of receiving more ways. At regular intervals, this information is collected by a centralized agent and ways are re-allocated among cores. Much of the StimulusCache design [55] is subsumed by the CloudCache design [54]. The StimulusCache design was motivated by the observation that cores are much more likely to emerge from the manufacturing process with faults than cache banks. The design attempts to allocate the "excess" cache banks belonging to faulty cores to other functional cores. In addition to suffering from the usual complexity of a private LLC (finding an LLC block via a directory), these designs also suffer from the added complexity imposed by D-NUCA (block search and block migration). As described earlier, the work of Cho and Jin [41], Awasthi et al. [42], and Chaudhuri [45] attempt to similarly allocate cache space among competing cores in the context of a shared LLC.

MorphCache, Srikantaiah et al., HPCA'11

Similar to the CloudCache, the MorphCache design [56] also starts with a private cache organization and allows reconfiguration to form larger caches. Depending on an analysis of working set requirements, private L2 or L3 cache slices are either merged to form a larger shared cache slice or previously merged slices are split into smaller private slices. Thus, each cache slice (either L2 or L3) may either be a private cache or it may be a part of a larger cache that is shared by 2-16 cores. If L2 cache slices are merged, the corresponding L3 cache slices are also merged. Merging is triggered when there is imbalance in the utilization of adjacent slices or if highly-utilized slices are dealing with the same shared data. A merged cache slice has a higher UCA latency. As discussed next, an optimized shared cache baseline will often offer the nice features provided by many of the individual options in a reconfigurable private cache design.

2.3   DISCUSSION

We compared the properties of generic shared and private caches in Table 1.1. Most recent papers on private cache organizations continue to cite the disadvantages of generic shared caches while advocating the private cache organization. Such arguments are often mis-leading because they turn a blind eye to the several innovations in recent years that have largely alleviated the primary disadvantages of a shared cache. For example, several papers mention that shared caches are inefficient because they incur high access latencies on average and because there is higher load on the network used to access the shared cache. These arguments apply to earlier shared cache designs that employed set-interleaved mapping or that implemented a physically contiguous shared cache. They no longer apply to recently proposed designs. Even a simple design that employs a distributed tiled shared S-NUCA cache, page-to-bank mapping, and first-touch OS page coloring goes a long way towards ensuring that most LLC requests do not leave the local tile.

Similarly, papers on shared caches often highlight a private cache's poor hit rate while not giving due credit to Cooperative Caching techniques. Also, the benefit of easy data look-up in a shared LLC is only possible if the LLC is inclusive.

That said, a private cache organization likely has a steeper hill to climb in terms of overcoming its pitfalls. A major pitfall is the complexity involved in locating a line that may be resident in other private caches. Private caches have been used in commercial implementations with few cores; the use of a broadcast bus facilitates finding data. However, this approach does not scale well as the number of cores increases. An on-chip directory would be required and is not easy to maintain. The key difference is this: in a shared S-NUCA cache, we know exactly where to find a block; in a private cache organization, we need to "search" for the block because it could be anywhere. Private caches will of course continue to be common in modern multi-cores; they will be most effective if they are eventually backed up by an on-chip shared inclusive last-level cache that can handle coherence among the private caches. Cooperation will likely not be required for such private caches because evicted lines will anyway be found in the shared LLC. We therefore believe that innovations for shared caches, of the kind discussed in Section 2.1 and the next chapter, will likely have more long-term impact.

CHAPTER 3

Policies Impacting Cache Hit Rates

The previous chapter focused on policies that blurred the line between shared and private caches, and that attempted to bring data closer to requesting cores. Most of the papers discussed in this chapter are oblivious/agnostic to the non-uniform nature of cache access latencies. Therefore, much of the discussion here will focus on hit rates. For the most part, this chapter will assume a shared LLC and examine various policies for block insertion, replacement, and fetch. What distinguishes this work from the vast literature on caching from prior decades is the emphasis on multi-thread or multi-program working sets. Part of the work can be broadly classified as *"cache partitioning"* and the metrics of interest are usually cache hit rates and quality-of-service (QoS). While Chapter 2 discusses some papers [20, 22, 26, 27, 41, 42, 46, 47, 52] that partially fall under the cache partitioning umbrella, that chapter was more focused on a collection of non-uniform latency banks (either private or shared) and on data placement policies to improve proximity. We will continue to mention some of these previously discussed papers to clarify their contribution to the cache partitioning literature.

3.1 CACHE PARTITIONING FOR THROUGHPUT AND QUALITY-OF-SERVICE

3.1.1 INTRODUCTION

The Baseline Platform

Most papers on cache partitioning attempt to split the ways of a set-associative monolithic shared cache among multiple competing threads or programs. This is a reasonable model for small-scale multi-cores. As discussed earlier, a large-scale multi-core will likely be tiled, and a shared LLC will likely be distributed, say, one cache bank per core. When a shared cache is distributed, we can either distribute the sets, or the ways, or both. If ways are distributed, a given cache line could reside in one of a number of banks and block search mechanisms will be required. This is the complex D-NUCA model. As argued in Chapter 2, a more scalable model employs an S-NUCA architecture with OS-based page coloring to improve data proximity. S-NUCA employs set-partitioning, and all ways of a set are typically contained in a single bank. Cache partitioning can be accomplished by allocating sets to cores with miss-rate aware page coloring [41, 42]. This was discussed in Section 2.1.3 and will also be briefly summarized at the end of this section. With such a model, a given bank typically houses part of a large working set for an application or a small working set for one application and the spilled working sets for a few other large applications. No one has yet studied way-partitioning (or set-partitioning) for the latter situation and its interaction with the page coloring policy. It is not clear if the expected benefit will be high. Even if we assumed that complex D-NUCA was the way of the future, way-partitioning would have to be aware of the non-uniform latencies to different ways, and most cache partitioning papers do not consider this ([57] being the only exception).

Therefore, most of the work in this section has applicability in small-scale systems where a shared monolithic LLC is employed or in a large-scale system where each tiled bank of the LLC is shared by multiple cores. In both cases, we expect a small number of cores to compete for the available cache space.

Policies within the Cache Replacement Policy

Capacity in the cache is typically adjusted during cache replacement. Traditionally, the LRU block replacement method is used in most caches. This requires the maintenance of a recency stack for each set to indicate the ordering of the last access to each block in that set. The replacement process involves the following three primary policies:

1. *Victim Selection:* Traditionally, this has been the LRU block in the recency stack for that set.

2. *Block Insertion:* When a new block is inserted, it is traditionally inserted as the MRU block in the recency stack.

3. *Block Promotion:* When a block is touched, it is moved to the MRU position in the recency stack.

Most cache partitioning papers attempt to deviate from the baseline LRU policy by altering one or more of the above three policies.

Optimization Strategies

It is also worth noting that cache partitioning can have multiple reasonable performance targets. Hsu et al. [58] show that the optimal partition varies significantly depending on the metric being optimized. They identify three major types of performance targets: (i) *Communist*, that ensures that each application achieves a similar level of performance or performance degradation; (ii) *Utilitarian*, that tries to maximize overall throughput by assigning resources to threads that can best utilize the additional cache space; and (iii) *Capitalist*, that refers to an unregulated competition for resources, similar to the baseline LRU policy. Hsu et al. show that Capitalist policies can lead to partitions that are highly sub-optimal in terms of overall throughput and fairness. They also show a good general correlation between the other two performance targets, *i.e.*, with a few exceptions, a policy that optimizes fairness also provides adequate overall throughput (and vice versa). The outlier exceptions make it necessary to identify the performance target when designing a cache partitioning policy. Section 3.1.2 primarily focuses on Utilitarian policies, while Section 3.1.3 focuses on Communist policies.

Clever cache management obviously has little to offer if application working sets fit comfortably in cache. These policies are important when the total working set exceeds the available cache space. In such a scenario, the policies have to decide what blocks get to stay in cache and what blocks are evicted. Prior work has always assumed LRU for cache replacement, which is a Capitalist policy. Behavior with LRU is governed by the rate at which applications issue data requests; if an application accesses lots of data, it is allocated lots of cache space. But this additional cache space need not necessarily help reduce the cache miss rate for this demanding application. Instead, it may hurt other co-scheduled applications that may be robbed of the little cache space they need to enjoy high hit rates. This is the main insight behind the recent policies that advocate a shift from Capitalist to Utilitarian or Communist policies: that space be allocated based on the expected benefit or impact, and not simply based on activity levels.

Dynamic Set Sampling

This chapter introduces several innovative policies for cache management. Rarely does a single policy emerge as a clear winner on every benchmark program. These policies also usually work best if their parameters are optimized for each program. A common feature in many papers is the use of a mechanism referred to as *dynamic set sampling* or *set dueling monitors (SDMs)* to help each program identify optimal policies and parameters. This was a technique introduced by Qureshi et al. [53, 59]. In a large cache with several sets, a few sets can be used to monitor the behavior of a given policy or parameter. For example, if the cache has 4096 sets, 16 sets can be hardwired to use *policy-A*, and 16 sets can be hardwired to use *policy-B*. Miss counters can track the number of misses in these 16-set groups and decide if *policy-A* is better than *policy-B*. The better policy can then be applied to the other 4064 sets of the cache. Thus, less than 1% of the cache space is used for experimentation; even if some policy in this "experimental lab" is highly sub-optimal, extra misses incurred here should be in the noise. The main benefit of course is that an optimal policy can be selected for each program at run time. When dealing with multi-core workloads, the number of possible options can sometimes grow dramatically, and this may require the use of many more set samples.

3.1.2   THROUGHPUT

We start by describing the few bodies of work (UCP, TADIP, PIPP, $AGGRESSOR_{pr} - VT$) that are currently considered the state-of-the-art in throughput-based cache partitioning. We then briefly describe some of the prior work that provided initial insight in this area. UCP performs explicit partitioning by allocating cache ways to applications based on the marginal utility provided by each way. TADIP and PIPP build on recent work that claim that insertion policy innovations are more effective at constructing an optimal cache population. They therefore strive for implicit cache partitions by offering each thread different insertion options, a concept that will be shortly expanded upon. While these schemes are explicit or implicit forms of way partitioning, some papers have also looked at explicit set partitioning, an approach that in some cases is readily implementable in modern hardware with page coloring [46].

UCP, Qureshi and Patt, MICRO'06

Cache partitioning across competing threads is done by the LRU replacement algorithm, by default. Qureshi and Patt [60] make the observation that LRU is oblivious of the marginal utility that more cache space can provide to a given thread. Their Utility-based Cache Partitioning (UCP) mechanism estimates the marginal utility of each way to each thread and selects a partition that minimizes cache miss rates. The key contribution is the design of a low-overhead monitoring mechanism (UMON) that efficiently computes the necessary marginal utilities.

Each core has a UMON unit that stores tags for the L2 cache and manages these tags as if the L2 cache is only being accessed by this core. These "shadow" tags are managed with an LRU replacement policy. If a request hits in (say) the 7th most-recently used way of the shadow tags, the request would have been a hit only if the cache partitioning policy allocated at least seven ways to this core. A counter tracking the marginal utility of the 7th way is incremented. One such counter is maintained for each way for each UMON unit. Every epoch (say, 5 million cycles), these counters are examined, a way partition is selected to minimize overall miss rates, and the counters are reset (or halved to retain some history). In order for the replacement policy to enforce a way partition within a set, each block must have a few-bit tag to indicate its "owner" core. A block is selected for eviction in a manner that enforces the per-core way quotas. It is fair to say that UCP primarily alters the victim selection policy: once a victim core is selected, the baseline LRU victim selection, insertion, and promotion policies are applied within the ways assigned to that core.

Note that the selected way partition applies to the entire cache, and way partition decisions are not made for each individual set. The shadow tag overhead can be reduced with sampling; for example, the monitoring can be done for every 32nd set. This brings down the storage overhead of the UMON unit to only a few kilo-bytes. But each core needs a UMON unit, so the overhead scales up linearly with the number of cores. Storage overhead can be further reduced by maintaining partial tags.

While the way-partitioning decision is trivial for two competing cores, the number of possibilities that must be considered increase exponentially as more cores compete. As clarified earlier, perhaps this is not a major concern as a monolithic cache unit will likely only be shared by a few cores. Qureshi and Patt also suggest heuristics for quick decision-making [60]. They start with a standard greedy algorithm and refine it so the algorithm not only examines marginal utilities for the next way, but it is aware of upcoming (in the algorithm) big jumps in marginal utilities. The decision making process is also complicated if accesses are made to both shared and private data blocks, but this is not considered in [60].

Some of the initial insight for marginal utility based cache partitioning was provided by the work of Suh et al. [61] (discussed shortly). Other work, described in the previous chapter, have also allocated ways to each core [26, 27] and used run-time metrics to estimate the marginal utility of additional ways [27]. A couple of recent papers have pointed out that small improvements can be made to UCP if the optimization function focused on maximizing core IPC instead of minimizing overall MPKI [62, 63].

TADIP, Jaleel et al., PACT'08

Caches typically employ LRU replacement policies with the incoming block occupying the MRU position in the access recency stack. To reduce confusion, we refer to the baseline block insertion policy as *MRU Insertion Policy (MIP)*. In a paper in ISCA'07, Qureshi et al. [53] hypothesize that such a policy is incompetent at retaining relevant blocks for many workloads (workloads that are streaming or have very large working set sizes). For such applications, it is more effective to mark the incoming block as LRU instead of MRU. This is referred to as the *LRU Insertion Policy (LIP)*. Thus, the insertion policy is being changed, but the victim selection and block promotion policies are the same as the baseline LRU mechanism. A refinement on this policy is *Bimodal Insertion Policy (BIP)* that with a small probability places a block in the MRU position. For a given application, the optimal insertion policy (BIP or MIP) is selected with a *Set Dueling Monitor (SDM)* that employs each policy on a few sets and determines the best performing one (see Section 3.1.1). This is referred to as the *Dynamic Insertion Policy (DIP)* and is discussed in more detail in Section 3.2.1. Figure 3.1 pictorially represents many of the recent insertion policies.

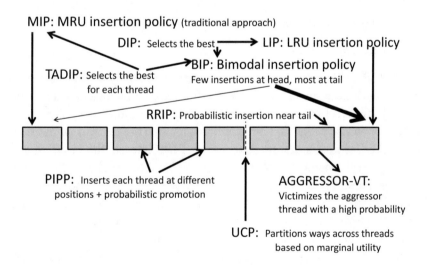

Figure 3.1: Depiction of various insertion, promotion, and victim selection policies.

Jaleel et al. [64] first show that the use of DIP within a shared cache is effective. In this case, either BIP or MIP is used for all applications sharing the cache. The authors then propose *Thread Aware DIP (TADIP)* that makes a choice of BIP or MIP for each individual thread. By selecting the appropriate insertion policy for each thread, the lifetime of each cache line is influenced and, accordingly, a cache partition is achieved. The notion of marginal utility (as in the UCP paper) is somewhat encoded within TADIP because it determines the threads that do not benefit from keeping a block around. These threads are forced to occupy fewer ways by inserting their blocks in the LRU position. But TADIP is more effective than UCP because even applications with large working sets get to keep part of their working set (the part with highest locality) in cache because of BIP's probabilistic occasional insertion into the MRU position.

Unlike UCP, TADIP does not require shadow tags to estimate the optimal cache partition. TADIP requires that we estimate the optimal insertion policy (BIP or MIP) for each thread. This is again done with SDMs with two different strategies. The first strategy, referred to as *TADIP-Isolated*, attempts to figure out the optimal policy for each thread while assuming that the other threads employ MIP. One SDM is used to monitor behavior for sets that use MIP for all threads. N other SDMs are used, where each SDM uses BIP for one thread and MIP for all others (N is the number of cores or threads). A second strategy, *TADIP-Feedback*, attempts to figure out the optimal policy for each thread while being aware of the policy selected by the other threads. A total of $2N$ SDMs are used, where each thread uses two SDMs. One of these two SDMs uses BIP for the "owner" thread and the other uses MIP; for all other threads, both SDMs use the current optimal policy as selected by other SDMs. The total extra storage overhead of these policies is quite small. But it is important to ensure that the sets forming the SDMs are a small fraction of the total sets in cache, *i.e.*, only a moderate number of cores can be supported.

PIPP, Xie and Loh, ISCA'09

The TADIP approach uses a variable insertion policy to achieve a pseudo-partition of the cache; pseudo, because there is no strict specification or enforcement of a partition. In their ISCA'09 paper [65], Xie and Loh introduce a modification to the insertion policy. Further, they suggest an alternative *promotion policy*. The combination is referred to as *Promotion/Insertion Pseudo Partitioning (PIPP)*. TADIP impacts the insertion policy; PIPP impacts the insertion and promotion policy; neither impacts the victim selection policy. Note that these baseline policies refer to the position of a block in the recency stack. With new policies in place, the stack is not necessarily ordered by access recency. Therefore, the victim selection process is evicting a block at the tail end of what used to traditionally be the recency stack; this block is not necessarily LRU. Hence, it is more appropriate to refer to the stack as a priority stack than as a recency stack.

Instead of inserting a block at either MRU (high-priority) or LRU (low-priority) positions (as is done by TADIP), PIPP allows insertion at an arbitrary position in the stack. If an application is deemed to benefit from higher capacity and is granted a large cache partition, all of its blocks are inserted in a higher-priority position in the stack. The determination of this position is done by the OS, possibly with input from a UMON circuit that estimates marginal utility curves for each core.

Instead of immediately moving a block to a high-priority position upon touch, PIPP allows the block to move up the stack by a single position with a specified probability.

It is easy to see that there may be scenarios favoring TADIP or PIPP. If a block is only touched twice in quick succession, TADIP will promote the block to the high-priority position and unnecessarily prolong the block's cache lifetime. On the other hand, if a block is only touched once, TADIP is quicker in evicting the block. To better deal with the latter scenario, Xie and Loh use more aggressive policies when streaming or thrashing workloads are detected – blocks for such workloads are inserted at low-priority positions and have lower probabilities for promotion. By using the probabilistic BIP policy, TADIP can retain a portion of a thrashing workload, while PIPP uniformly inserts every block of an application in the selected position. Overall, the results of Xie and Loh show that PIPP is superior to TADIP and UCP.

In addition to the PIPP policy, Xie and Loh propose a novel monitoring circuit, referred to as an *In-Cache Estimation Monitor (ICEmon)*. Similar to set dueling, a few sets are designated to monitor behavior for each core. For those sets, traditional LRU is used to manage the blocks of their assigned core. Blocks from cores other than the assigned core are only allowed to occupy low-priority positions in the stack. Counters keep track of the marginal utility afforded to the assigned core by each additional cache way. These estimations are noisy once the low-priority ways are considered, but this noise is not relevant. Such a monitor does away with the need for sampled shadow tags in each core.

$AGGRESSOR_{pr} - VT$, Liu and Yeung, PACT'09

Recent work by Liu and Yeung [66] keeps the insertion and promotion policies intact (same as baseline LRU) but, similar to UCP, focuses on the victim selection policy (which was untouched by TADIP and PIPP). They first make the observation that an aggressor cache-thrashing thread is selected for victimization 80% of the time by an oracle replacement policy that looks into the future to minimize miss rates. This leads to their proposed replacement policy: (i) if an aggressor thread happens to be the LRU block in the set, it is victimized; (ii) if not, the aggressor thread's LRU block is victimized with a 99% probability. This strongly biases the replacement policy to evict the aggressor thread's blocks. They also make the observation that for many threads and workloads, overall throughput (in terms of weighted IPC) is maximized when the aggressor thread uses a victimization probability of either 99% or 50%.

Therefore, at run-time, the following procedure is adopted (referred to as $AGGRESSOR_{pr} - VT$). Similar to the algorithm in [67], measurements are taken every epoch to detect a phase change and a phase change triggers an exploration mode to identify the policy to be used by each thread. Each thread is considered sequentially and executes with other threads for an epoch each as a non-aggressor and as an aggressor with victimization probabilities of 50% and 99%. Of the many combinations considered, the one with highest throughput is employed until the next phase change is detected.

Interestingly, for their workloads, the authors show little room for improvement over a baseline LRU policy. IPC and miss rate improvements over baseline LRU are only around 5-6% even with the oracle replacement policy. Within this small space, the authors show that the $AGGRESSOR_{pr} - VT$ policy is clearly better than UCP. While the mechanisms are different, this policy strives for the same behavior as TADIP and PIPP – all three try to limit the space occupied by thrashing workloads by either inserting them in low-priority positions or favoring them for victim selection.

Suh et al., Journal of Supercomputing, 2004

A part of the basic UCP framework, including the marginal utility approach and the mechanisms for cache partitioning, were initially proposed by Suh et al. [61]. The major contribution of UCP was the low-overhead estimation of marginal utility for each core with sampled shadow tags. Suh et al. estimated marginal utility with counters per way and even counters per group of sets. LRU information was also maintained to track recency of access among sets. By combining these counters, it is possible to estimate the marginal utilities of additional cache space at finer granularities than an entire way. The OS takes this into account and specifies partition sizes for each process; this is considered when making replacement decisions. Each cache block tag must now store the id of its owner thread, and there are counters to track the number of blocks in the cache per thread. While UCP's shadow tags are not required here, the marginal utility estimation either requires that an application execute in isolation or risk a spurious estimation because of interference from other applications. Dropsho et al. [68] also employed counters per way to estimate miss rate curves for their *Accounting Cache*. This was done to reduce energy by disabling ways, not to partition the cache.

Yeh and Reinman, CASES 2005

Yeh and Reinman proposed an adaptive shared cache partitioning scheme [57] that had some features that were later also incorporated into other partitioning schemes. For example, they proposed the use of shadow tags to quickly estimate an application's miss rate curve at run-time, a feature also seen in follow-up work [60]. An epoch-based scheme is used to allocate ways to applications every epoch, with a greedy algorithm that maximizes throughput, while ensuring that no application is significantly worse than if it had its own private cache bank. Unlike most work in this area, Yeh and Reinman consider a D-NUCA cache and assign ways to cores keeping proximity in mind. Like all D-NUCA schemes, the design suffers from the complexity of search and migration.

Rafique et al., PACT'06

Rafique et al. [69] propose hardware *mechanisms* and OS *policies* to allocate shared cache capacity across competing threads (the shared cache has uniform access times). Similar to the work of Suh et al. [61], there are mechanisms to enforce per-process quotas in terms of ways or overall cache space. The latter quota is a little more fair in its capacity allocation but may do a poor job of handling some hot sets. The OS employs various policies to determine quotas and sets these by writing to special hardware registers. The OS policies can either employ a static allocation or can monitor performance counters and try to equalize miss rates per thread or match per-thread IPCs to per-thread priorities.

MTP, Chang and Sohi, ICS'07

Chang and Sohi [70] make the observation that for many workloads, a slight increase in allocated capacity is far more beneficial than the degradation caused by a decrease in allocated capacity. For these workloads, it is therefore better to implement an unfair partition at any time, where one application receives most cache resources and other applications receive a small fraction of resources. The other applications attempt to recoup some of their lost performance in later epochs when each receives most of the cache space. Chang and Sohi point out that cycling through these biased partitions is better for throughput and QoS than using the same partition throughout execution. The mechanism is referred to as *Multiple Time-Sharing Partitions (MTP)*. The authors also combine MTP with their Cooperative Caching scheme [48] so that the better of the two policies is used every epoch, depending on the workload.

Adaptive SET PINNING, Srikantaiah et al., ASPLOS'08

The work of Srikantaiah et al. [71] is not quite intended as a cache partitioning mechanism, but it does serve to allocate LLC space among competing cores. It is primarily designed to improve overall cache hit rates by reducing cache interference. It does so by allocating sets among cores (unlike most other papers that directly or indirectly allocate ways among cores).

The paper first introduces the notion of intra- and inter-processor misses. This specifies if a block was evicted by the same processor (intra) that accesses it next or by a different processor (inter). This is clearly an approximate classification. It is a sampling of who happens to push a block off the edge of the LRU stack. But it does not encode the events that push the block to the brink of eviction.

The paper argues that a few hot cache blocks are responsible for most inter-processor misses, and it is best to segregate these blocks in a cache region that is private to the accessing core. Each core therefore maintains a *Processor Owned Private (POP)* cache that is considered part of the L2 level of the hierarchy but has a size similar to typical L1 caches. In addition, every set in the L2 is assigned to the first core that touches it. This achieves some level of cache partitioning based on first access to a set. If a core brings in a block that indexes to a set belonging to another core, it is forced to place the block in its own POP cache. Similarly, when looking up a block, not only must we look up the set that the block indexes to, we must also look up all the POP caches. The entire L2 cache is thus free of intra-set interference.

The mechanism is referred to as *Set Pinning*. To prevent sub-optimal cache partitions because of what may have transpired during the first touches to sets, processors are forced to relinquish their ownership of sets during run-time. Each set maintains a counter that keep tracks of whether the set is yielding more misses for non-owner cores than hits. If that is the case, the owner core gives up ownership for that set. This is referred to as *Adaptive Set Pinning*.

Set-partitioning can also be achieved with alternative schemes such as page coloring, and it has been employed within NUCA caches [41, 42] and UCA caches [72]. This is discussed at the end of this section.

ACCESS, Jiang et al., HPCA 2011

In a recent HPCA'11 paper, Jiang et al. [73] argue that future multi-cores will likely employ a collection of heterogeneous private LLCs, where each LLC may itself be shared by multiple cores. Given the diversity of workloads on future multi-cores, the use of different sized LLCs prevents over-provisioning and improves performance per watt. Jiang et al. describe the hardware and OS support required to compute a mapping of threads to resources that optimizes throughput on such an architecture.

Each private LLC maintains shadow tags for each core that shares the LLC and each LLC associativity found in the system. Set sampling is used to greatly reduce the size of these shadow tags. This structure allows the OS to estimate the miss rate if each thread executed by itself on each LLC. This information is gathered at the start of each program phase. In order to compute the miss rate of a given schedule, the OS performs various calculations. It first estimates an equation for the miss rate curve, based on the few data points it has collected from the shadow tags. It then estimates a cache allocation per core based on the relative miss rates of the threads sharing that private LLC in the schedule being considered. The above two estimations are used to compute the overall miss rate for that schedule. Since the number of possible schedules can be very high, performance estimations are only made when tasks enter/exit or when a thread enters a new program phase. When this happens, the OS considers a few incremental schedules that vary based on how the new thread is accommodated. Ultimately, the architecture shows how the overall LLC cache space can be partitioned across many cores in a manner that is close to optimal.

Software Techniques for Better Cache Sharing

Most cache partitioning work has assumed the existence of specialized hardware to estimate the optimal partition. For example, the UCP work of Qureshi and Patt [60] assumes shadow tags to estimate the miss rate curve (MRC) of a program as a function of allocated cache size (associativity). More recent work, such as TADIP [64], does not require estimation of MRCs, but it assumes extra hardware for set dueling monitors (SDMs). As a result, while these ideas are great candidates for upcoming processors, much of this work cannot be supported in existing processors.

While the hardware-based techniques have assumed a pre-specified workload that shares a cache, it should also be possible for the OS to construct run-time workloads or program schedules that can minimize (but can likely not eliminate) inter-thread interference. For example, Chandra et al. [74] describe two heuristic models and one analytical model that can predict the degree of interference if provided with the stack distance profile of each thread as input. Assuming that such information is available, the OS can first make sure that co-scheduled threads are as minimally disruptive to each other as possible; the hardware-based policies described previously would then be exercised to further reduce the disruption. We next describe two recent OS-based techniques that can be implemented in modern processors to improve sharing behavior in caches and possibly complement other hardware-based techniques.

Tam et al. [75] show how MRCs can be estimated at acceptable cost in modern processors, thus enabling the OS to implement UCP-like schemes. On the Power5 processor, Tam et al. record the address of each L1 data cache miss in a register and raise an exception. The invoked system call then records the register value in a trace log in memory. The key is that hardware performance monitoring units are being leveraged and costly binary instrumentation is avoided. But because of the frequent exceptions, this process also incurs non-trivial overheads, causing a 4X application slowdown. However, the process must only be invoked at the start of the program or the start of each program phase. An adequate trace can be recorded in a few hundred milli-seconds. A standard stack distance algorithm can then be employed to analyze the trace and estimate the MRC (again consuming a few hundred milli-seconds). Since the trace contains L1 data cache misses, it is independent of other threads that may be scheduled on the other cores. Tam et al. use the MRCs generated at run-time to implement a page coloring policy [41] that partitions sets among cores to minimize overall miss rate. Some of the same authors also go on to use page coloring based techniques to isolate problematic pages to a small region of the LLC. This work is described in more detail in Section 3.2.1.

Zhuravlev et al. [76] focus on practical OS scheduling mechanisms to improve cache sharing. They focus on a processor model that has multiple shared caches and attempt to assign applications to cores such that overall cache interference is minimized. Their proposed approach will likely also apply to other processor models that have a single shared cache. They make the observation that a simple examination of miss rates of applications is enough to achieve a good classification of applications. They make the case that more complex approaches with stack distance profiles are not required. The miss rates for each application can be sampled at run-time and even the noise introduced by other co-scheduled applications at the time of sampling does not result in poor scheduling decisions. Once these approximated miss rates are gathered, applications are sorted based on miss rates, and applications assigned to cores such that the overall miss rates of each shared cache are roughly equalized. This is referred to as the *Distributed Intensity Online (DIO)* mechanism.

Set Partitioning Schemes

While the last chapter focused on bridging the gap between shared and private caches and reducing latency in a NUCA cache, many of those papers were also partitioning the LLC across many cores, either implicitly or explicitly. For example, Qureshi's Dynamic Spill-Receive design [52] designated private LLCs as spillers and receivers, thus implicitly partitioning the LLC into unequal ever-changing quotas. Beckmann et al. [20] made that partition slightly more explicit by introducing hardware cost/benefit measures and varying the probabilities for replica creation.

Of the work described in the previous chapter, the ones that represent the most explicit form of cache partitioning are those of Cho and Jin [41], Lin et al. [46, 47], and Awasthi et al. [42]. All of these studies, and even that of Tam et al. [75], assume a shared LLC and partition sets among cores. The partition is effected with page coloring: forcing each core to allocate data in pages that map to sets assigned to that core. This puts a greater onus on the software; Lin et al. [46] and

Tam et al. [75] even show that such set-based cache partitioning is realizable in modern hardware. In contrast, the explicit and implicit way partitioning mechanisms discussed in this section require additional hardware support. When considering a NUCA cache, way partitioning is compatible with a D-NUCA design, while set partitioning is compatible with the simpler S-NUCA design. As described in the previous chapter, set partitioning can be made more explicit by introducing cost functions to determine the preferred color for every page [42], and partitions can be dynamically altered by migrating pages at a significant hardware cost [42, 45].

3.1.3   QOS POLICIES

The previous sub-section focused on *Utilitarian* policies to maximize overall throughput. As described by Hsu et al. [58], other reasonable performance goals exist, such as the *Communist* policies that attempt to equalize performance or performance degradation for all co-scheduled threads. Similarly, policies can combine throughput and fairness, for example, Yeh and Reinman's attempt to maximize throughput while capping the degradation of each thread [57]. A variation of the *Communist* policy is an *Elitist* [77] policy that places different priorities or different performance degradation constraints on each application. The above policies are required if the system is expected to provide certain guarantees to each running application, as might be the case in datacenters or other platforms that "rent out" computing power to clients. This sub-section describes the mechanisms and recent innovations that help provide such quality-of-service (QoS) guarantees.

Iyer et al., SIGMETRICS 2007, ICS 2004

Most prior work on QoS has attempted to schedule CPU time for applications in order to meet performance constraints. In a multi-core CPU, many applications are executing simultaneously and sharing resources such as the LLC and memory bandwidth. Iyer et al. [77] show that policies to allocate cache space and memory bandwidth can be very helpful in tuning the performance of individual applications and allowing each application to meet its performance constraints. The paper lays out the basic framework for a best-effort QoS-aware platform. If guarantees stronger than best-effort must be provided, then the system must have the capability of activating more servers when the currently active servers cannot meet all performance guarantees.

For starters, a *minimum performance constraint* must be specified for every application (the performance constraint for a high-priority application is also frequently referred to as the *target*). Multiple metrics can measure the efficacy of the QoS policy, but application execution time or throughput are the metrics that are typically employed. The QoS policy can be either *static*, where resource quotas for each application are specified beforehand, or *dynamic*, where performance constraints are specified beforehand and the hardware/OS is in charge of monitoring behavior and adjusting resource quotas at run-time. Resource quotas influence the shared LLC space allocated to each application as well as the memory bandwidth available to each application. Memory bandwidth allocation is an essential part of the QoS system to prevent priority inversion, *i.e.*, a low-priority application with a small cache quota could experience several cache misses and negatively impact the memory bus availability for the high-priority application.

Iyer et al.'s QoS-aware memory architecture has three layers. The first is *priority classification* that requires administrators to specify priorities and constraints. The OS is responsible for propagating this information to hardware QoS registers on every application context-switch. The second layer is *priority assignment,* which requires the conversion of OS specified constraints to actual resource quotas based on performance monitoring. The quotas are saved in a hardware QoS resource table. Each cache access issued by the CPU is accordingly tagged as high or low priority before it is sent to the LLC. The third layer is *priority enforcement* and requires counters to track memory system and cache space usage. The cache quota enforcement requires a replacement policy that is aware of the usage of each application. The enforcement of memory bandwidth allocation requires that a high-priority application be allowed to issue N memory requests (if present in the memory controller queue) before relinquishing the memory bus for other lower-priority requests.

In prior work, Iyer [78] focused on the problem of *priority enforcement* and suggested different mechanisms for the LLC. The first mechanism involves way partitioning, either reserving specific ways for an application (referred to as static) or imposing a cap on the number of ways that an application can occupy (referred to as dynamic). The second mechanism attemtps to control cache space per application by probabilistically choosing to either cache a fetched line or not. The third mechanism implements different cache structures (fully-associative or direct-mapped caches, stream buffers, victim caches, etc.) and maps them to applications based on need and priority. Individual cache lines are also allowed to mark themselves, so they are always cached or immediately self-invalidated, based on priority. In their follow-on work, Iyer et al. have focused on a priority enforcement scheme that is based on tracking overall cache space usage of each application.

In more follow-on work, Varadarajan et al. [79] advocate organizing the cache into many small (8-32 KB) direct-mapped banks called *molecules.* This enables fine-grain cache resource allocation for QoS. The cache is partitioned across multiple cores at the granularity of molecules. This allows high flexibility and low power; although, the paper ignores the interconnect overhead in accessing many small banks (an aspect considered later in detail by Muralimanohar et al. [7]). The molecular cache approach allows full flexibility in organizing molecules to form a core's cache, *i.e.,* some sets could have more molecules (and hence higher associativity) than other sets. For many applications, such high flexibility is likely not worth the implementation cost.

Guo et al., MICRO 2007

Iyer et al.'s SIGMETRICS'07 paper was followed by a MICRO'07 study [80] that afforded more flexibility to the QoS manager. The authors first argue for the following guidelines. QoS requirements are better specified in terms of easily quantifiable resources, such that it should be easy to verify the availability of the requested resources. In other words, a QoS target specification in terms of IPC or miss rate makes the QoS manager's job harder, but a QoS target specification in terms of cache capacity is easy to handle. If a server does not have the requested resources for a task, it is stalled (or it must be assigned to an idle server). Guo et al. also argue that tasks should be classified in different ways: it should be possible for users to specify that their tasks can afford some performance degradation (referred to as the *Elastic execution mode*), perhaps incurring a lower service fee in return.

Having the above flexibility makes it easier to maximize the overall throughput on a given server. If every application made strict resource requests, resource fragmentation can happen. For example, a server may have some spare resources that never get used because each new task demands more resources than what is available on the server. Each task may also be utilizing fewer resources than what is specified in the strict request. Tasks that use the Elastic execution mode may be able to relinquish some of their resources to help accommodate more tasks on the server and boost throughput. This can be implemented in two ways. A new (otherwise inadmissible) task can downgrade to the Elastic execution mode and try to make do with whatever resources happen to be available on the server. Alternatively, resources can be stolen from other tasks on the server without violating the specified guarantees. Resource stealing would require mechanisms such as sampled shadow tags [60] to detect any violations.

In another follow-up paper, Zhao et al. [81] articulate the cache book-keeping structures (*CacheScouts*) that would be required in future processors. In essence, it is important to understand how much cache space is occupied by a process, and how processes interfere in the cache (either constructively or destructively). This helps the QoS manager, helps the OS make smart scheduling decisions, helps programmers tune applications, and helps service providers charge clients for resource usage. In order to compute the above metrics, process ID tags must be associated with each cache line. Overheads can be reduced primarily by sampling only a few sets. Counter arrays are required to estimate interference by observing the process IDs of incoming and evicted blocks.

SHARP, Srikantaiah et al., MICRO 2009

Srikantaiah et al. [82] bring a more formal control theoretic approach to the basic framework of Guo et al. [80]. Mathematical formulations are used to express IPC as a function of cache miss rate, allowing IPC to be specified as a performance target (unlike the recommendation of Guo et al.). The framework also minimizes oscillations in the allocation of cache ways to applications. If the required ways are more or less than the available ways, ways are re-allocated to either maximize throughput or preserve assigned priorities. While the control theory framework can guarantee maximal throughput when the required ways are less than the available ways, only best-effort QoS can be provided if the required ways exceed the available ways.

Srikantaiah et al., Supercomputing 2009

While multi-core QoS papers have focused on cache and memory bandwidth partitioning in tandem (for example, [77]), the work of Srikantaiah et al. [83] was the first to consider cache and processor partitioning. If the multi-core workload consists of multiple multi-threaded applications, throughput is a strong function of the number of cores assigned to each application as well as the shared cache resources assigned to each application. There is a vast search space of processor and cache partitions that must be considered. Srikantaiah et al. prune the search space by dynamically constructing regression-based models that express IPC as a function of processor count and cache resources. This is used to iteratively predict optimal partitions, and the partitions are adjusted as more configurations are sampled and the regression models are refined. The partitions can be selected to either optimize throughput or provide QoS guarantees.

Virtual Private Caches, Nesbit et al., ISCA 2007

Nesbit et al. [84] focus on mechanisms required for the QoS-aware partitioning of bandwidth into the shared cache. The mechanisms are similar to those used in networking. The shared cache maintains a buffer of pending requests per thread. If a thread has been allocated half the bandwidth share and a shared cache access takes 20 cycles, it is expected that the request will take 40 cycles. A per-thread register is set to indicate that the thread cannot issue its next request until the 40 cycles have elapsed. When the next request from that thread shows up (before the 40th cycle), its expected completion time is set to 80 cycles. If the next request shows up after the 40th cycle, the expected completion time is set to 40 cycles after the request arrival time. Thus, the expected completion time for each thread's next request is computed, and the scheduling order is based on earliest estimated completion time. The above policy gracefully allocates any excess spare bandwidth to threads that have minimally utilized cache bandwidth to that point. It also of course ensures that a thread is provided roughly its promised share of cache bandwidth. Within a thread's buffer, re-orderings are permitted, *e.g.*, reads may be prioritized over writes. Higher-level OS/feedback mechanisms are required to estimate the cache bandwidth share that is appropriate for each thread. Without such a QoS-aware scheduling policy, it is easy for a cache-intensive application to starve other threads with a baseline FIFO scheduling policy.

Kim et al., PACT 2004

In early work, Kim et al. [85] studied the problem of fair cache partitioning, a close cousin of the QoS problem. In fair cache partitioning, an attempt is made to uniformly degrade all co-scheduled applications, where degradation is measured relative to each application's performance when executing in isolation. Kim et al. specify a number of reasonable target metrics based on cache misses to estimate "performance". They introduce static way-partitioning policies that require prior knowledge of each application's stack distance profile. They also introduce a dynamic epoch-based scheme that iteratively identifies the most and least impacted applications and allocates a way from the least to most impacted application (in an effort to equalize the negative impact on all co-scheduled applications). If this re-allocation does not change performance much, it is cancelled after the next epoch. Kim et al. show that optimizing for fairness often results in overall throughput improvements, while the reverse is not true, *i.e.*, a policy that focuses on throughput optimization can often lead to unfairness.

3.2 SELECTING A HIGHLY USEFUL POPULATION FOR A LARGE SHARED CACHE

This section focuses on policies that decide what to bring in to the cache, what to keep, and what to evict. These are attempts to maximize cache efficiency by doing more with less. We begin by discussing replacement/insertion policies that decide what data to keep when an application suffers from many capacity misses. We then discuss optimizations for associativity, largely targeted at conflict misses. Finally, we look at block-level optimizations that include compression, prefetch, and elimination of dead/useless blocks.

3.2.1 REPLACEMENT/INSERTION POLICIES

While replacement policies have been well-studied in the past [86, 87, 88, 89], the emergence of large shared LLCs has led to a flurry of recent observations and innovations in this area. It has been generally believed that some variation of the LRU policy is best at retaining relevant blocks in the cache. However, accesses to multiple data structures from multiple applications, each filtered by the L1 caches, are multiplexed into the LLC. Depending on the nature of these accesses, the use of LRU can be highly sub-optimal, and this is the premise behind much of the recent work covered in this sub-section.

LRU assumes that a recent touch to a block is the best indicator of its use in the near future. However, several blocks do not fit this description. Some access patterns are best described as *scans* [90], where a large number of blocks are touched once and are re-referenced again in the distant future. This happens, for example, when sequentially reading the contents of a large array. A *thrashing* access pattern [90] involves a collection of blocks that do exhibit re-use, but the re-use distance exceeds the size of the cache. Both access patterns and their mixtures would behave very poorly with LRU. Further, when using a non-inclusive hierarchy, blocks with high temporal locality may remain in L1 or L2 and the LLC version of the block is typically only touched once. For all of these access patterns, it is important to identify blocks that have very long resue distances and not retain them in the LLC. A common theme in some of the work, described shortly, is this notion that a block must "prove its worth" before it is retained. This is often done by modifying the insertion policy. Another common theme is the prioritization of blocks based on frequency of use. Yet another body of work is based on the observation that half the blocks in cache at any time have already serviced their last use and need to somehow be identified as dead blocks. Some of this dead block prediction work is covered in Section 3.2.3.2.

LIN, Qureshi et al., ISCA 2006

In an ISCA'06 paper, Qureshi et al. [59] argue that the replacement policy should not just take recency into account, but also the *"cost"* of fetching a block from memory. This cost is determined by examining the *memory-level parallelism (MLP)* at the time of a miss. In essence, a cache miss that happens in isolation has a high cost, while a cache miss that happens at the same time as other cache misses has a low cost. This notion is similar to the notion of cache block criticality that has been explored previously [91, 92], but Qureshi et al. use new metrics that explicitly focus on MLP. There is also prior work on integrating cost and recency in the replacement policy [88], although that work has focused on varying cost because of NUMA.

Qureshi et al. first introduce counters per entry in the MSHR that are incremented every cycle based on the number of active entries in the MSHR. Once a miss is serviced, the MSHR entry counter provides an estimate of the cost incurred in servicing that miss, referred to as *mlp-cost*. It is assumed that the next miss for that block will incur similar mlp-cost, and, accordingly, the block is deemed more or less precious. This estimate is stored in the tag entry for that block. The implementation thus has the nice property that the mlp-cost of the block is estimated when the block is fetched and a separate predictor is not required. The replacement policy uses a linear combination of the block's recency and its mlp-cost, and it evicts the block with the lowest such combination. Hence, the policy is referred to as *Linear (LIN)*. However, the policy sometimes performs worse than LRU, especially if the mlp-cost is not consistent across successive misses for a block. Qureshi et al. therefore employ a tournament-style predictor that dynamically decides whether to use LRU or LIN. A sample of sets determines the cost of misses when using LRU for that sample, and another sample of sets determines the cost of misses when using LIN for that sample. The better policy is used for the rest of the cache. This initial incarnation of Set-Dueling is referred to as Sampling Based Adaptive Replacement (SBAR).

While the mlp-cost approach has merit and relatively low implementation cost, it has received relatively little attention in recent years. We believe that it may be worthwhile to examine the interaction of the mlp-cost metric with other more recent replacement policies.

Adaptive Caches, Subramanian et al., MICRO 2006

Subramanian et al. [93] generalize Qureshi et al.'s idea of adaptively selecting the better of two replacement policies. In their work, they focus on selecting the better of LRU and LFU on a per-set basis. This is done by having separate auxiliary tag structures that track hypothetical cache contents with either replacement policy. A bit vector per set also tracks recent history on which policy may have yielded a hit while the other may have yielded a miss. The use of such a sliding bit vector provides some theoretical bounds on performance that may not be possible with saturating counter based implementations. On every eviction, the better replacement policy for that set is used to determine the block that must be evicted. Because we may frequently switch between policies for a set, the main tag storage for the cache may not resemble either of the auxiliary tag storages. The overhead for the auxiliary tags may be reduced by using partial tags or by doing set sampling, without a significant impact on performance. The use of set sampling would require that a uniform policy be applied for all sets in the cache; although, this too seemed to have little effect on performance for the studied workloads.

DIP, Qureshi et al., ISCA 2007

We described the DIP technique of Qureshi et al. [53] in Section 3.1.2 and re-visit that discussion here. Since the baseline block insertion policy in any large shared cached places an incoming block into the MRU position of the access recency stack, we refer to it as the *MRU Insertion Policy (MIP)*. In many workloads, it is better to insert a block into the LRU position (*LRU Insertion Policy, LIP*). This gives a block a brief stint within the cache and is an appropriate approach when dealing with streaming blocks. If the block is touched again before its eviction, it is moved to the MRU position and stays in the cache for a longer time. In a *Bimodal Insertion Policy (BIP)*, most blocks are placed in the LRU position, and with a small probability, an incoming block may be placed in the MRU position. This is effective at retaining a relatively "hot" portion of a large working set within the cache.

Ultimately, Qureshi et al. advocate a *Dynamic Insertion Policy (DIP)* that employs either BIP or MIP for most of the cache. This decision is made by using the BIP policy on a few sets and the MIP policy on a few other sets and using the better policy on all other sets. Such use of set samples to pick among competing policies is referred to as a *Set Dueling Monitor* (see Section 3.1.1). Only 32 or 64 sets need to be used for these samples and misses in these sets increment or decrement a single saturating counter. Thus, apart from the logic used to adaptively manage the recency stack and keep track of the sets that constitute the SDM, there is almost no additional storage overhead. The work of Qureshi et al. therefore offers new insight to a well-studied problem and an approach that is a significant advancement without being overly complex.

Reuse Distance Prediction, Keramidas et al., ICCD 2007

A replacement policy can approximate Belady's OPT replacement algorithm [86] if it evicts a block that is touched furthest in the future. Keramidas et al. [94] attempt to make high confidence reuse distance predictions for blocks. The replacement policy uses a combination of LRU and reuse distance information. Among blocks that have reuse distance predictions, they select the block that is accessed furthest in the future. Among blocks with no reuse distance predictions, they select the block that was accessed furthest in the past (the LRU block). The block that is furthest away in either its next predicted access or its previous access is selected for eviction. Keramidas et al. use an instruction-PC based predictor for reuse distance. When a block is fetched into the L2, the corresponding instruction's PC is used to index into CAM and RAM structures that may yield a high confidence reuse distance prediction for the block. Multiple prediction structures must be navigated when updating and looking up the predictor.

Shepherd Cache, Rajan and Govindarajan, MICRO 2007

In their MICRO 2007 paper, Rajan and Govindarajan [95] make an interesting observation. They show that a cache with half the associativity, but an optimal replacement policy (Belady's OPT [86]) can out-perform a baseline LRU-based cache. Based on that observation, they devise a cache organization where the cache is organized as two logical partitions (way partitions), and one of the partitions is used to collect information that allows us to emulate OPT on the other partition.

One of the partitions is referred to as a *Shepherd Cache (SC)* and uses a FIFO replacement policy. In most experiments, the Shepherd Cache has 4 ways and the *Main Cache (MC)* has 12 ways. On every cache miss, the incoming block, A, is placed in the SC. It remains in the SC through the next three cache misses in that set; at which point, the block A is at the head of the FIFO queue for the SC. On the next cache miss, we must determine if the block A should be evicted or if it should be retained at the expense of a block in the MC. We are attempting to manage the MC with a replacement policy that is OPT-like. In some sense, the decision of who to evict from the MC when A is brought in should have been made many cycles ago, but it has been deferred until this point. By deferring the decision, we have been able to examine what happens beyond the fetch of A; this look into the "future" allows us to make an OPT-like replacement policy within the MC. At this point, either A or a block from the MC is evicted. This block is chosen by examining the accesses that happened since the fetch of A; we either pick the block that was accessed last since the fetch of A (exactly matching the replacement decision that OPT would have made for the MC), or in case of multiple candidates, fall back on an LRU policy. The organization requires a few complex structures to keep track of accesses since A was fetched and how those accesses impact the selection of replacement candidates.

An alternative view to this policy is this: every incoming block is forced to stick around for at least four other evictions; at that point, we make a decision to either keep the block or not, based on recent behavior. The Shepherd Cache organization may therefore be somewhat analogous to a PIPP-like policy [65] (Section 3.1.2) where a block is inserted four places away from the tail of the priority list.

Extra-Lightweight Shepherd Cache, Zebchuk et al., ICCD 2008

Zebchuk et al. [96] attempted a practical implementation of the Shepherd Cache philosophy. As in the original Shepherd Cache design [95], incoming blocks are placed in a few FIFO ways designated as the Shepherd Cache (SC). Instead of tracking the orders of next accesses to blocks [95], Zebchuk et al. simply track if blocks in the SC have been touched or not. When a block A graduates to the head of the SC FIFO and if it has not been touched, it is evicted. If the block A has been touched, then the LRU block in the main cache is evicted, and A is inserted in the LRU position of the main cache. This implementation has low complexity and makes the design appear more like the alternative perspective mentioned in the previous paragraph. Zebchuk et al. also reduce complexity by considering practical baseline replacement policies that approximate LRU. They point out that practical implementations of pseudo LRU [97] can yield miss rates that are 9% higher than true LRU.

RRIP, Jaleel et al., ISCA 2010

In an ISCA 2010 paper, Jaleel et al. [90] present a replacement policy that has low complexity and that combines the insight from a large body of prior work. We first describe their observations and some of this prior insight before delving into their proposed design.

For almost any replacement policy, the blocks in a set of the LLC are organized as a priority list. The assumption is that a high priority block is likely to be re-referenced sooner than a low priority block. Hence, when we place a block within the priority list, we are essentially making an implicit *Re-Reference Interval Prediction (RRIP)*. In LRU, we assume that an incoming block or a recently accessed block has a *near-immediate* or very short re-reference interval. Another baseline policy used in modern processors because of its simplicity is *Not Recently Used (NRU)*. In this policy, a single bit is associated with each block. Upon every touch, the bit is set to 0 to indicate recent use. If a block must be evicted, we simply pick the first encountered block that has its bit set to 1. If all blocks have their bit set to 0, all bits are set to 1. Because only a single bit is used to indicate recency, it is difficult to discriminate between the priorities of the many blocks in the set. A primary contribution of Jaleel et al. is the use of multiple bits per block to track priorities at a finer granularity and the mechanisms to initialize and update these priorities on subsequent accesses.

The RRIP policy of Jaleel et al. [90] is most similar to NRU, but it tracks information at a finer granularity. Every block has an M-bit counter associated with it (the design defaults to NRU for $M = 1$). The counter value is an implicit measure of the re-reference interval prediction: a low value implies that we expect the block to be re-referenced in the near future, and a high value implies that we expect the block to be touched in the distant future. In the *Static-RRIP (SRRIP)* policy, an incoming block's counter is initialized to $2^M - 2$, one less than the counter's maximum value of $2^M - 1$. A block is selected for eviction only if it's counter value is $2^M - 1$. If no such block exists, the counters for all blocks in that set are incremented until a counter saturates and a block can be evicted. When a block is touched, its counter can be decremented by one (this is the RRIP-Frequency Priority policy and closely resembles LFU [98]) or directly set to zero (the RRIP-Hit Priority policy that is closer to LRU). The authors show that the latter performs better because a single re-reference shows that the block is not part of a scan (streaming access with no immediate reuse) and should be retained. In addition to the just described Static RRIP, the authors also introduce Bimodal RRIP, where an incoming block's counter is initialized to either $2^M - 1$ or (with a low probability) $2^M - 2$. Further, SDMs can be used to dynamically select the better of Static and Bimodal RRIP, thus yielding the *Dynamic RRIP (DRRIP)* policy.

The primary complexity introduced by RRIP is the maintenance of M-bit counters for every block in the cache. Corresponding operations for look-up, comparison, increment, and decrement are required. This complexity is higher than that required for NRU and slightly lower than that required for a dynamic policy that selects between LRU and LFU. The SRRIP policy requires less hardware than LRU, but it outperforms LRU.

In essence, RRIP out-performs DIP because incoming blocks are being given more time to determine their re-use potential. While DIP places most incoming blocks at the tail of the priority list, RRIP places most incoming blocks at an intermediate position in the list by initializing the block's counter to a non-maximum value. This bears similarity with the philosophy of the Shepherd Cache [95]. The RRIP policy can also be viewed as a generalization of NRU, combined with an intermediate insertion policy. As Xie and Loh [65] point out, there are unlimited possibilities in terms of how blocks can be inserted and promoted on the priority list and to some extent, the work of Jaleel et al. tries to capture the best traits of many successful prior policies (DIP, LFU, NRU, Shepherd Cache). The two main types of access patterns that create problems are *scans* and *thrashes*. Scans are single accesses to a stream of data with no re-reference in the near future. Thrashes are accesses to large portions of data with a large enough re-use distance that blocks are evicted before their re-use. These access patterns are especially problematic when they interleave with other access patterns that do exhibit temporal locality. The family of replacement policies (DIP, PIPP, RRIP, etc.) that insert blocks close to the tail of the priority list are primarily targeting tolerance to scans and thrashes, with subtle variations in the extent of tolerance.

Pseudo-LIFO, Chaudhuri, MICRO 2009

In his MICRO 2009 paper, Chaudhuri [99] introduces the novel concept of a fill stack. A *fill stack* keeps track of all blocks in a set in the order in which they arrived. A given block will initially reside at the top of the stack and gradually descend down until it is evicted from the cache (perhaps before it reaches the bottom of the stack). Given this somewhat predictable traversal of a block through the fill stack, Chaudhuri makes the argument that a block's position in the fill stack can be used to estimate its liveness. Chaudhuri shows that most cache hits are serviced by blocks at the top of the fill stack. In other words, blocks even at the top of the fill stack have likely served their utility and should be prioritized for victimization. This motivates a family of replacement policies that leverage fill stack information and that believe in the priority of *last-in-first-out (LIFO)*. Even when most hits are not localized to the top of stack, it is easy to identify "knees in the hit curve" – positions in the fill stack beyond which there is a sharp drop in blocks that yield hits.

Before the fill stack information can be used, the application is first executed with traditional LRU at the start of every phase and information is gathered on how fill stack positions yield hits on average. This information is gathered every epoch and is also used to detect phase changes. Three "knees" in this hit curve are identified – fill stack positions beyond which there is a drop-off in blocks that yield hits. Three policies are considered, one for each knee. In each policy, a block is victimized if (i) it has not yet served a hit in its current fill stack position and (ii) it is closest to the top of the fill stack while being above the knee position. A set dueling mechanism is used to select the best of these three policies and LRU. The primary overhead of this *probabilistic escape LIFO (peLIFO)* policy is that each block must now track its current fill stack position (must be updated on every eviction), and the fill stack position that yielded its last hit (so that the hit curve can be estimated).

Chaudhuri also suggests two alternative policies. One uses a dead block predictor and uses the fill stack to break ties if there are multiple candidate dead blocks. The second is based on the premise that if (say) 75% of hits are yielded beyond fill stack position X and 25% of hits are yielded beyond fill stack position Y, then 50% of blocks should be evicted between positions X and Y. Based on these estimated probabilities, a block in a given fill stack position is selected for victimization.

Unlike DIP and RRIP that only hasten the eviction of a block that is touched exactly once, Chaudhuri's peLIFO policies can also hasten the eviction of blocks that are touched multiple times. In that sense, they are most closely related to prior work on dead block prediction but incur significantly lower overhead. The notion of dead block prediction has been exploited for various cache replacement policies [100, 101, 102] and is discussed next in Section 3.2.3.2. In Chaudhuri's design, the notion of a fill stack is also used to exploit distant temporal locality. Additional hits are caused by the blocks that descend to the bottom of the fill stack, are retained in a region of the LLC that can be viewed as a "victim cache", and are able to yield hits in the distant future.

Scavenger, Basu et al., MICRO 2007

Chaudhuri's MICRO'09 paper on peLIFO [99] identifies the existence of long-term reuse. The peLIFO policies probabilistically retain some blocks that might be temporally dead, and these blocks later yield additional hits. However, peLIFO does not explicitly try to retain blocks that are likely to yield hits in the distant future. Some prior work by Basu et al. [103] attempts to do just that.

Basu et al. partition their LLC into two large regions – the regular L2 and a *victim file (VF)*. When a block is evicted from L2, it is not indiscriminately placed in the VF, unlike a traditional victim cache [104]. The block is placed in the VF only if it is likely to yield hits in the future. A counting Bloom filter is used to estimate the number of misses for this block, and this metric is used to determine future reuse and the block's priority for retention in the VF. A min-heap is used to efficiently track the lowest-priority block in the VF; another block is retained in the VF only if it has a higher priority than this lowest-priority block. The large VF is organized as a direct-mapped hash table with a doubly-linked list to maintain all blocks that map to the same hash. While these innovations help reduce the complexity of maintaining a large victim file, there is little precedence in terms of commercial processors implementing pointer-based hardware data structures. There may well be more room for improvement in defining a complexity-effective technique to retain blocks with distant reuse.

NUcache, Manikantan et al., HPCA 2011

Manikantan et al. [105] propose a novel method to identify useful data that should be retained in the cache to yield hits in the distant future. First, they identify the PCs that yield the most cache misses. These PCs are referred to as *delinquent PCs*. The ways of the LLC are partitioned as *MainWays* and *DeliWays*. All incoming blocks are placed in the MainWays that are managed with an LRU replacement policy. A block evicted from the MainWays is retained in the DeliWays if it was brought in by a set of chosen delinquent PC. The DeliWays follow a FIFO replacement policy. The selection of "chosen delinquent PCs" is best explained with an example. If a delinquent PC was the only one allowed to use 4 DeliWays and accounted for 50% of all misses, a block would be inserted into the DeliWays at every other miss. The block would be evicted out of the DeliWays by the time the set experiences 8 misses. Hence, a block's *next-use distance* can tell us if keeping the block around in the DeliWays could have been useful or not. The next-use distance for a block is defined as the number of misses experienced by that set between the eviction of the block and its next use. If the histogram of next-use distances for blocks brought in by a delinquent PC is known, we can estimate the benefit of adding the delinquent PC to the chosen set. The cost of adding this delinquent PC is that it pushes other blocks sooner out of the DeliWays. These calculations are non-trivial, and the chosen set is constructed greedily by incrementally selecting the delinquent PC with highest benefit.

The most delinquent PCs are identified by keeping track of the number of misses yielded by recent PCs. Incoming blocks are kept in another table, and the global miss count is recorded in the table when that block is evicted. When the block is touched again, we compute the next-use distance. This updates the next-use histogram for the corresponding delinquent PC.

Pollute Buffer, Soares et al., MICRO 2008

Soares et al. [106] propose a software-based page coloring scheme to improve hit rates. They argue that it is best to isolate pages with high miss rates to a small region of the LLC. Such pages are deemed to be polluters because retaining them in cache does not lead to many hits for their cache blocks, and they, in turn, evict other useful blocks. The authors therefore designate $1/16^{th}$ of the LLC as a pollute buffer, and OS page coloring steers polluter pages to be mapped to sets in the pollute buffer. This serves as a form of intra-application cache partitioning for higher hit rates.

Soares et al. use hardware performance counters and a Linux kernel module to track various L2 hit/miss statistics for each virtual page at run-time. After initial profiling at the start of program phases, the most cache-unfriendly pages (the ones with highest L2 miss rates) gradually undergo page migration. An exploration phase copies different numbers of pages until an optimal assignment of pages to the pollute buffer is identified.

Managing Inclusion, Jaleel et al., MICRO 2010

The trade-offs between inclusive and non-inclusive caches are well-known. Inclusive caches simplify cache coherence because tag look-ups must only be performed at the larger inclusive cache. However, inclusive caches waste space (especially when the inclusive cache is not much bigger than the previous levels of the hierarchy) and result in *"inclusion victims"*. Inclusion victims are blocks that have high temporal locality and are present in L1 caches; because the L1s filter out these accesses, these blocks drift to the LRU position in L2, are evicted, and are also removed from L1 to preserve inclusion. While these problems and corresponding solutions have been known for decades, Jaleel et al.'s recent study [25] again focuses a spotlight on this problem in the modern multi-core context. This is especially important because nearly every recent paper on LLC replacement policies has focused on a non-inclusive hierarchy and has largely ignored the role of inclusion victims in inclusive hierarchies. Future papers also need to clearly specify how inclusion is managed in their simulator; many recent papers do not even specify if the L1-L2 hierarchy is inclusive or not.

Jaleel et al. show that while the difference between inclusion and non-inclusion is not significant for single-thread workloads, it is significant for multi-core processors and workloads. They also show that the difference between inclusion and non-inclusion can be attributed more to inclusion victims than the difference in effective cache hierarchy capacity. If the LLC is eight times larger than the previous level of the hierarchy, non-inclusion does not improve performance by much; the difference is more stark when the LLC is only four times larger or less. This matches the design choices in recent Intel (inclusive) [107] and AMD processors (non-inclusive) [108]. When inclusion does suffer a performance penalty, that penalty can be eliminated with a replacement policy that checks with the L1 before L2 eviction (assuming an L1-L2 inclusive hierarchy). In an inclusive hierarchy, when a block is evicted from L2, the L2 anyway sends messages to L1 to evict the block from L1s. If the block is found in L1, instead of evicting the block, the L1 responds with a NACK that forces the L2 to find another victim. This process can typically be hidden behind the latency to fetch the incoming block from memory.

3.2.2   NOVEL ORGANIZATIONS FOR ASSOCIATIVITY

The papers discussed in this sub-section make the observation that some sets are more pressured than others and hence vulnerable to more conflict misses. They make an attempt to equalize load across sets with hardware-based schemes.

Before delving into these papers, we describe a few other alternatives that are worth considering. The problem of load imbalance among sets may be alleviated with previously proposed hardware/software techniques. For example, page coloring policies can reduce the non-uniformity in accesses to sets, thus obviating the need for more complex hardware. Such an approach was proposed by Sherwood et al. [72] to reduce misses in large caches. When sets are determined to be hot, a recently touched page is re-assigned from those sets to a different color. The TLB tracks this on-chip re-naming of a virtual page. Indexing functions have also been proposed to balance activity across sets [109, 110]. Ultimately, the easiest way to reduce conflict misses is to increase associativity. Since LLCs typically employ serial tag and data access, higher associativity only impacts the energy for tag access.

V-Way Cache, Qureshi et al., ISCA 2005

In Section 2.1, we had discussed the NuRAPID organization of Chishti et al. [24]. The NuRAPID design was attempting to yield low access latencies in a NUCA cache by affording flexible data placement in cache banks. To make this happen, the data and tag stores were decoupled, *i.e.*, a tag entry was allowed to point to any data block in any bank, instead of to a fixed entry in the data store. Managing such a cache requires forward and reverse pointers in the tag and data stores to match the tag and data for a block. A similar inspiration was used by Qureshi et al. [28] in their ISCA'05 paper to improve hit rates in an LLC.

In conventional caches, a block maps to a unique set, and the fixed collection of data-store entries reserved for that set. Qureshi et al. argue that flexibility in LLC data placement is good for two reasons: (i) it helps deal with an imbalance in load among cache sets, and (ii) it allows the use of novel replacement policies that are cognizant of data usage beyond a single set. The flexibility is provided with the following organization.

The tag-store is allowed to accommodate more entries than the data-store. It is similar in organization to a conventional tag-store in that it has a fixed number of sets and ways; an important distinction is that each tag-store entry maintains a forward pointer to indicate the data-store entry corresponding to that tag. If the tag-store has twice as many entries as the data-store, half the tag entries will be invalid. In some sense, the tags are over-provisioned, allowing a given set to accommodate twice as many entries as its fair share. This helps absorb some variation in load per set. The data-store is simply a collection of cache blocks with no direct fixed correspondence to tag-store entries. Every data-store entry must maintain a reverse pointer to its corresponding tag-store entry.

When a block is brought in, if no free tag entry exists for that set, LRU is used to evict one of the tag entries in that set. The incoming block is placed in the data store entry corresponding to the tag just evicted. When a block is brought in, if a free tag entry exists for that set (this is true most of the time because of the over-provisioning in the tag store), we must next identify a data block to replace from the data store. Because of the existence of forward and reverse pointers, the incoming block can replace any block in the data store. The replacement policy need not just examine access recency within that set, it can evict any block in the cache that is likely to have already served its use. This global replacement policy is implemented as follows. Every data store entry has a 2-bit saturating counter that is incremented on every access. The replacement engine scans through these counters looking for a zero counter value; the first encountered block with a zero counter value is evicted. During this scan, every examined counter value is decremented. When handling the next replacement, the scan starts where it previously left off.

This new organization leads to better replacement decisions and handles load imbalance among sets. The primary penalties are a larger tag store, the maintenance of forward and reverse pointers, and the logic and saturating counters used by the replacement engine.

Set Balancing Cache, Rolan et al., MICRO 2009

Rolan et al. [111] make the argument that sets in a large cache experience non-uniform activity and pressure. When this happens, sets that experience high pressure are associated with sets that experience low pressure. Pressure is measured with saturating counters associated with each set that are incremented/decremented on misses/hits. The associations can be static (sets are associated if they only differ in the most significant bit of their indices) or dynamic (a set with high pressure is associated with a set with least pressure). When sets are associated, the block evicted from the high-pressure set is placed as the MRU block in the associated set. When such associations are employed, block look-up must be performed sequentially in both associated sets. Such a set-balancing policy dynamically offers higher associativity for a few sets. Prior work in the area has focused on direct-mapped baseline caches, such as the *Balanced Cache* work of Zhang [112]. Recently, Khan et al. [102]

extend the Set Balancing Cache by including the notion of dead block prediction (discussed in more detail in Section 3.2.3.2).

STEM, Zhan et al., MICRO 2010

The Set Balancing Cache just described was also extended in another recent paper by Zhan et al. [113]. First, they introduce a more accurate (but more costly) measure to determine if sets will benefit from being associated. Each set in the tag array is augmented with partial (hashed) tags of recently evicted blocks. Saturating counters keep track of hits in these victim tags and determine if the set should spill or receive blocks. The saturating counters are also checked on every miss to confirm that the receiving set is not being overwhelmed. Second, the Set Balancing idea is combined with replacement policy innovations. The main tags and the victim tags use different replacement policies (either LRU or BIP [53]), and the better policy is used for each set.

The ZCache, Sanchez and Kozyrakis, MICRO 2010

Sanchez and Kozyrakis decouple the notion of "ways" and "associativity" in their *ZCache* design [29]. Ways represent the number of tags that must be searched when looking for a block. They refer to associativity as the number of blocks that could be evicted to make room for an incoming block. The ZCache keeps the number of ways small, but it has a large associativity. However, a significant downside is that blocks have to be copied and re-arranged on most cache misses.

The ZCache builds on Seznec's skewed-associative cache design [109]. In a skewed-associative design, each way of the cache uses a different indexing (hashing) function. This leads to fewer conflicts than a traditional set-associative design. If the cache has four ways, a block A can reside in four fixed locations; when block A is brought into cache, it replaces a block in one of those four locations. The ZCache extends this replacement policy. The blocks in those four locations can potentially reside in other locations that do not conflict with block A. So, for example, block A could take the place of block B and B could be moved to one of its other three possible locations. Similarly, the blocks in those three locations could also be moved elsewhere. In other words, if we allow blocks to be re-located to one of their alternative locations, a tree of replacement possibilities opens up. By examining this tree up to a certain depth, we can find a replacement candidate that is most suitable. Sanchez and Kozyrakis use coarse-grained timestamps on blocks to select the LRU block among several candidates. The incoming block then causes a series of copies within the cache until the LRU block is finally evicted. While the replacement process is not on any latency-critical path, it makes the ZCache more expensive in terms of hit and miss energy than a baseline with the same number of ways. Overall, energy can be reduced if the better replacement decisions can significantly reduce the number of off-chip accesses.

3.2.3   BLOCK-LEVEL OPTIMIZATIONS

This section discusses papers that attempt to optimize the quality of blocks brought into the cache. This is roughly organized as: (i) papers dealing with block prefetch, (ii) papers dealing with dead block prediction, and (iii) papers identifying useless words or performing compression. These topics also relate strongly with the topic of cache replacement policies.

Prefetching has been actively studied for at least three decades, with significant contributions made by Chen and Baer [114], Jouppi [104], and Smith [115]. Prefetching can be applied to most levels of the hierarchy; it is most effective when prefetching blocks into the LLC since off-chip latencies can often not be tolerated with out-of-order execution. The switch to multi-core has not been a major game-changer for the prefetching arena although some papers have reported a drop in prefetch effectiveness in multi-cores [116], and some work [117] has considered the co-ordination of multiple prefetchers on a multi-core. Prefetching papers continue to battle the classic problems of accuracy, timeliness, pollution, and memory access contention. Prefetching is a vast enough area that we can't do justice to all recent work in this book – we will shortly discuss a sample that represents the state-of-the-art.

The prefetching concept has also been connected with the concept of dead block prediction. Dead block prediction was first introduced by Lai et al. [118] (and was also considered by Wood et al. [119]). In that work, they show that the end of a block's utility can be predicted by constructing a trace of all instructions that access that block. Once a block is considered dead, that acts as a trigger to evict the block and prefetch a useful block in its place. Hu et al. [120] also use dead block prediction as a trigger for prefetch, but instead of constructing traces, they use a period of inactivity to deem a block dead. Both of these works are focused on the L1 cache. In this sub-section, we describe more recent advancements in this arena that have focused on dead block prediction for the L2 cache. Dead block prediction has been used for prefetching [101, 118, 120, 121], cache bypassing [100, 101], cache replacement policies [100, 101, 102], and for energy reduction [122, 123, 124].

Finally, this section also discusses papers that have examined the utility of words within a cache block and have attempted optimizations to either reduce energy or increase effective cache capacity. Effective cache capacity is also increased with orthogonal optimizations that compress a cache block.

3.2.3.1 Prefetch

Prefetching is an actively studied area and is often orthogonal to the organization of the cache hierarchy itself. Basic prefetch designs that continue to be heavily used today include Jouppi's stream buffer [104], Hu and Martonosi's Tag Correlated Prefetcher [125], and Nesbit et al.'s Global History Buffer [126]. This section covers just a representative sampling of recent work in the area. Interested readers are encouraged to see other recent papers on the topic [127, 128, 129]. It is also worthwhile to study the methodology and taxonomy introduced by Srinivasan et al. [130] to analyze the various effects of each prefetch operation.

Discontinuity Instruction Prefetcher, Spracklen et al., HPCA 2005

Most prefetching work has focused on data blocks. We present the work of Spracklen et al. [131] as a state-of-the-art example of an instruction prefetcher. That paper also provides a good summary of prior art in instruction prefetching, organized as sequential (stream-like), history-based (predictors), and execution-based (runahead-like) prefetching.

Spracklen et al. show that prior instruction prefetching approaches are not very effective for commercial workloads (databases, web servers) as they have large instruction working set sizes with many direct and indirect branches and function calls (accounting for 60% of all instruction cache misses). They argue that a low complexity predictor can be built if they focused on simply identifying discontinuities in instruction fetch, *i.e.*, fetches to non-contiguous cache lines. Every time this is observed, the discontinuity and the address of the fetched line are recorded in a predictor (if there is room). When fetching a cache line, the predictor is examined and if there is a hit, the predicted next line is fetched. The discontinuity predictor is employed in tandem with a sequential next-N-line predictor, so the sequential prefetches trigger look-ups of the discontinuity predictor, and when a non-contiguous line is prefetched, we continue to fetch sequential lines from the new target address. The predictor is direct-mapped; each entry only tracks last-time history. Each entry also has a 2-bit saturating counter; the counter is incremented on a useful prefetch and decremented if another discontinuity wants to use the same entry. Replacement can only happen when the saturating counter is zero.

Spracklen et al. observe that instruction prefetching is not as effective for the L2 because of useful data blocks that get evicted. Hence, prefetched instruction blocks are only placed in L1; they are placed in L2 only if they were useful before their eviction from L1. Every generated prefetch must also check cache tags before sending the request to the next level; this creates contention for cache tags and a large prefetch queue. Prefetches are therefore filtered before placement in the queue. Prefetches are dropped if the address matches a recent demand request. This and other simple optimizations allow 90% of all prefetch queue requests to eventually be sent to the next level.

Temporal Memory Streaming, Wenisch et al., ISCA 2005

Wenisch et al. [132] design a prefetching mechanism that is especially well-suited to multi-processor systems. They observe that consumers of shared data typically access the same sequence of addresses as previous consumers, referred to as *temporal address correlation*. Such a sequence of potentially arbitrary addresses is referred to as a *stream* (not to be confused with Jouppi's *Stream Buffer* [104] that refers to a sequence of contiguous addresses). Since these accesses to streams repeat frequently, the phenomenon is referred to as *temporal stream locality*. Hence, when a node i in a multiprocessor system has a miss, it checks to see if another node j recently had the same miss. It then prefetches the same sequence of addresses that j had subsequent misses on.

The implementation works as follows. Each node tracks its miss sequence in a large circular buffer that is stored in memory. When node j has a miss on block A, the address of A is recorded in the circular buffer at address X. Node j sends the address X to the directory, so it can be recorded as part of A's directory state. The directory can record multiple such pointers to streams. When another node i has a miss on block A, it contacts the directory and gets the latest value of A. It also receives information from the directory that a potential stream can be found by node j at address X. Node i

sends a stream request to node j, and node j returns the stream to i. Node i can eventually receive multiple possible streams that originate at address A. Node i then issues prefetches along these streams while they agree. Prefetched data is placed in a prefetch buffer. Upon disagreement, node i waits to see its own demand misses to resolve which stream is correct and then continues along that stream. When the stream is half consumed, node j is contacted again to receive the next entries in the stream. While there is a clear cost associated with storing streams in the circular buffer in memory, accesses to the streams are usually not on the critical path. More messages are introduced in the network as well. However, the benefits of prefetching can be substantial, especially in commercial applications. The notion of a stream is not restrictive: it is capable of handling sequential, irregular, and pointer-chasing data structures. In follow-up work, Wenisch et al. [133] add extensions to make the implementation more practical.

Ferdman et al. [134] also apply the concept of Temporal Streaming to the L1 instruction cache. Similar to Temporal data streaming, sequences of instruction cache misses are logged in memory (that gets cached in L2). The L2 tag for a block stores a pointer to the block's occurrence in the L2. This initiates a prefetch of subsequent blocks into the L1 instruction cache.

Spatial Memory Streaming, Somogyi et al., ISCA 2006

In their ISCA 2006 paper, Somogyi et al. [135] introduce the notion of a spatially correlated stream. In many applications, accesses within a region can be sparse, un-strided, but repeatable (and hence predictable). A bit vector can keep track of the cache blocks that are typically touched within a region. When a similar traversal over a region begins again, the bit vector guides the prefetch of the stream.

When a region (say, an operating system page) is touched, hardware structures keep track of the blocks touched within that region. The resulting bit vector constitutes the stream. The trigger (or starting point) for the stream is recorded as the PC of the first access to that region and the offset of the first block touched in that region. A stream for a region is ended when the cache evicts a block belonging to that region. The trigger and the bit vector are recorded in a predictor. On subsequent first misses to regions, the predictor is looked up; on a hit, the predicted bit vector is used to prefetch blocks in that region. This even allows us to accurately prefetch blocks from a region that may not have been previously touched, as long as the same PC performed a similar traversal on a different region.

Spatio-Temporal Memory Streaming (STeMS), Somogyi et al., ISCA 2009

Spatial memory streaming and temporal memory streaming target different groups of misses (with some overlap). The former can only identify misses to blocks within a single region (OS page). The latter can only make predictions for previously visited streams. Somogyi et al. [136] build a unified prefetching mechanism that can handle both types of access patterns. The temporal component of the predictor simply tracks the first access to each region and corresponding PC in its stream. These entries in the stream form the triggers that index into the spatial component of the predictor that predicts all the blocks that will be accessed within that region. This produces a flurry of blocks that must be fetched in the near future. So as to not overwhelm the memory system, prefetches to this large collection of blocks must be issued in an appropriate order. This requires that the temporal stream also record the number of misses between each entry; it also requires that the spatial predictor not just track a bit vector of accesses to that region, but the exact sequence of misses and the gap between consecutive misses to that region. With all this information, the exact expected sequence of misses can be reconstructed and timely prefetches issued.

Feedback Directed Prefetching, Srinath et al., HPCA 2007

Srinath et al. [137] introduce hardware mechanisms to estimate the accuracy, timeliness, and pollution effects of prefetches and use this information to throttle prefetch aggressiveness and dictate the management of prefetched blocks. They start with a baseline stream prefetcher that is modeled after that of the IBM Power4 processor [138]. The prefetcher is capable of handling multiple streams, and prefetch is initiated for a stream after detecting at least three misses in close proximity. The stream prefetcher attempts to stay P blocks ahead of a start address. Every time an item in the stream is touched, the start address is advanced by N blocks. P is referred to as the *prefetch distance*, and N is referred to as the *prefetch degree*. These two levers dictate how aggressive the prefetcher is. Srinath et al. employ five different levels of aggression by tuning the values of P and N.

Every interval, hardware metrics are examined to determine the prefetch accuracy, timeliness, and pollution. Accuracy is determined by storing a bit with every prefetched block to track its use. Timeliness is determined by seeing if prefetched blocks are accessed while they are still resident in the MSHR. Pollution is determined by keeping track of block addresses recently evicted by prefetches in a Bloom filter-style structure. Based on these metrics in the past interval and (to a limited extent) previous intervals, the prefetch aggression level is either incremented, decremented, or kept the same. The pollution metric is also used to modify the cache insertion policy. The prefetched block is placed in one of three locations in the latter half of the LRU priority stack based on the extent of pollution caused by prefetches. These schemes are effective and incur minor storage and complexity overheads.

A similar idea was also previously considered by Hur and Lin [139]. They too adjust the aggressiveness of a stream prefetcher, but this is based on hardware that estimates the typical lengths of streams and estimates if the current memory access is likely to be part of a longer stream or not. The strength of this technique lies in its ability to first issue and then curtail prefetches even for very short streams. The authors go on to improve the prefetch effectiveness by, among other things, biasing the stream detector to identify short streams [140].

Ebrahimi et al. [117] further leverage the notion of throttling of prefetchers based on dynamic feedback information. They first design a novel software-hardware co-operative prefetching mechanim for linked data structures. This is then combined with a stream prefetcher and run-time accuracy/coverage metrics are used to throttle whichever prefetcher is less effective. Competition for resources can not only happen with the use of multiple types of prefetchers, but also with prefetch streams for multiple cores of a processor. Ebrahimi et al. [141] again use dyanmic pollution/accuracy/bandwidth metrics to throttle prefetches for specific cores in an attempt to boost overall processor throughput.

3.2.3.2 Dead Blocks

IATAC, Abella et al., ACM TACO 2005

Abella et al. [123] attempt energy reduction in the L2 by turning off (gated-V_{DD}) cache lines that are predicted to be dead. A dead block prediction is made by considering a combination of time and block access counts. They first show that the average time between hits to a block is consistently less than the time between the last access and the block eviction. They also show that these times vary as a function of the number of times a block is accessed during its L2 residence. These observations motivate the following heuristic, *Inter Access Time per Access Count (IATAC)*, for dead block prediction and cache block turn-off. Every block keeps track of the number of accesses that have been made to it. Given this access count, a prediction is made for average time between hits. The prediction is based on global metrics for the entire cache and more heavily weighs the behavior of recently touched lines. Once the predicted time for the block elapses, the block is considered dead and is disabled. More than half the cache is disabled at any time with this policy, and there is a slight drop in performance because of additional misses created by premature block disabling.

Access Count Predictors, Kharbutli and Solihin, IEEE TOC 2008

Kharbutli and Solihin [100] build two dead block predictors that are based on counting accesses to cache blocks in the L2. Every block in L2 has an *event counter* that keeps track of accesses during the block's current residence in L2 and a *threshold* that conservatively represents the block's behavior during prior residences. When the event counter exceeds the threshold, the block is considered dead. It is then prioritized over other blocks by the cache replacement policy. The threshold estimation is also used to determine if the block is never accessed during its residence; in that case, the block is simply not cached in L2 on its next fetch if the L2 is pressured for space (referred to as cache bypassing in many prior works). The event counter can take two forms. The *Live-time Predictor (LvP)* tracks the number of accesses to a block during its L2 resisence. The *Access Interval Predictor (AIP)* tracks the number of accesses to the block's set between two successive accesses to the block. Past behavior for a block is stored in a table between the L2 and the next level of the hierarchy. This table is indexed with the PC of the instruction that causes the cache miss and the address of the missing block. The table entry stores the maximum event counter value previously seen by that entry and serves as the threshold prediction when a block is brought into L2.

Cache Bursts, Liu et al., MICRO 2008

Liu et al. [101] make the observation that counting individual references to a block can lead to inaccurate predictions, especially for irregular data access patterns. Hence, accesses are counted at the granularity of *Cache Bursts*. One cache burst is a series of references to a block during its continuous residence as the MRU block of a set. For example, during its L2 residence, a block may visit the MRU position of a set three times, servicing a number of accesses on each visit. After the last visit, the block moves from MRU to LRU position and is finally evicted. On the block's next L2 residence, when the block leaves the MRU position for the third time, it is predicted to be dead. Prior dead block predictors have also used the PCs of instructions accessing a block to form a trace [118]; now, the trace is formed with the PCs of instructions that cause the block to move into the MRU position. Liu et al. show that this new granularity for block access leads to higher accuracy and coverage for dead block prediction at the L1 and L2. The predictor is again placed between L2 and the next level of the hierarchy [100]. They also show that the moment that a block leaves the MRU position for the last time is also a good time to issue a prefetch that can replace the dead block. In addition to the prefetch trigger, the dead block prediction is also used to implement cache bypassing and prioritization in the cache replacement policy. The results of Liu et al. show that less than a third of the L2 cache contains useful blocks; this shows the vast room for improvement in L2 caching policies in general.

Virtual Victim Cache, Khan et al., PACT 2010

In recent work, Khan et al. [102] also show a similar room for L2 cache efficiency improvement. They show that a cache block is dead for 59% of its L2 residence time. Hence, the entire pool of predicted dead blocks in L2 is considered as a virtual victim cache that can house non-dead blocks that are evicted from the L2. Dead block prediction is done with Lai et al.'s trace-based predictor [118] with a few variations to reduce interference. To keep the implementation simple, only two sets are "partnered". When a block is evicted from one set, it is placed in its partner set. The partner set evicts a dead block or its LRU way. The evicted block is placed in either the MRU or LRU way of the partner set, depending on the outcome of a set-dueling implementation (see Section 3.1.1). Blocks are also prevented from ping-ponging between their home and partner sets. When looking for a block, the home set and the partner set are sequentially searched. The overall design is most similar to Rolan et al.'s Set Balancing cache [111], except that dead block prediction is used instead of identifying high and low pressure sets. The concept also has similarities to the Shepherd Cache [95] that also uses half the LLC as a victim file.

Sampling Dead Block Prediction, Khan et al., MICRO 2010

In a recent MICRO 2010 paper, Khan et al. [142] propose enhancements to the basic dead block predictor. Similar to prior work, the predictions are used for cache replacement and bypassing. They show that trace-based dead block predictions are not as accurate for an LLC in a 3-level hierarchy because of the filtering effect of the first two levels. Instead, high accuracy can be achieved by simply tracking the PC of the last instruction that touches the block during its LLC residence. The predictor complexity is reduced by keeping track of all prediction metadata in a separate tag array that maintains partial tags for a fraction of all LLC sets. This *sampler* structure is used to train a collection of saturating counters (the real predictor) that keep track of PCs that tend to generate last touches to blocks. In other words, access patterns for a few sets are used to identify the PCs that generate last touches for the entire LLC. When such a PC touches a block in the LLC, a bit associated with the block is set to indicate that it is dead. In fact, if the replacement policy often selects dead blocks for eviction, metadata for LRU can be eliminated from the LLC (although, it may be used within the smaller sampler structure). Thus, overall, the LLC design is greatly simplified and most complexity is localized to a small sampler structure.

3.2.3.3 Compression and Useless Words

On a cache miss, entire blocks are brought into the cache under the assumption that spatial locality exists in the access stream. However, in many workloads, only a fraction of the words (words are usually up to eight bytes wide) in a block are accessed and the rest are never touched. By only fetching the useful words in a cache block, (i) we can increase the effective useful capacity of the cache and (ii) reduce the bandwidth needs between the cache and the next level of the hierarchy. Similarly, compression of blocks is an orthogonal technique that can also be employed to achieve the same two benefits. There was some early work [143, 144, 145, 146, 147, 148] that examined prediction techniques to only fetch useful blocks, targeted primarily at L1 caches. Likewise, compression has been extensively studied for memory systems (for example, [149, 150, 151]) and for some L1 cache designs [152, 153]. We next describe recent work that applies these concepts to L2 caches.

Distill Cache, Qureshi et al., HPCA 2007

Qureshi et al. [154] devise a technique that, unlike prior work, improves performance without relying on predictors and without requiring selective fetch of blocks from the next level. On a cache miss, the entire block is fetched and placed in L2. During the cache's residence in L2, the utility of individual words in the block is learned. Once the block drifts close to the LRU position, they conclude that the words touched, so far, are useful and the rest are useless. The useful words are kept in cache while the useless words are evicted. This helps improve the effective capacity of the cache and boosts hit rate.

The L2 is now organized into two banks – a *line-organized cache (LOC)* and a *word-organized cache (WOC)*. The LOC resembles a traditional L2 cache; although, it has fewer ways. When a data block is evicted from the LOC, its useful words are moved to the WOC. The WOC has many ways, each way containing a single word; it therefore has high tag storage overhead. For workloads that exhibit a high degree of spatial locality, the authors introduce a scheme based on set-dueling monitors (see Section 3.1.1) to revert back to a traditional cache organization. The transfer from LOC to WOC only happens for half the lines evicted by L2 – those with a low percentage of useful words. The authors also attempt an optimization that performs compression on words in the WOC.

While the technique improves miss rates and performance, the implementation entails a fair bit of complexity; therefore, there is likely more research required to realize the potential from identifying useless words. The primary storage and energy overhead is introduced by the tag storage for the WOC. Since only the L1 caches see accesses to individual words, the L1 caches must track the utility of each word and merge that information into the L2 copy of the block. Additional bits are required in L1 and L2 to track word utility, and non-trivial interactions are introduced between the L2 and many L1s (especially if the L1-L2 hierarchy is non-inclusive). Copies are required between LOC and WOC and certain alignment constraints must be respected. Hits in the WOC result in partial blocks within the L1, requiring the use of a sectored L1 cache [146]. Reverting back to a traditional organization also requires extra tag storage.

Adaptive Cache Compression, Alameldeen and Wood, ISCA 2004

Data block compression introduces a fundamental capacity/latency trade-off: compressed blocks increase capacity, but they incur a higher read latency because they must be decompressed. To strike a balance in this trade-off, Alameldeen and Wood [155] first avoid compression in latency-critical L1 caches (similar to the approach of Lee et al. [156]). In the L2 cache, compression is employed for an incoming block only if the entire cache appears to be benefiting from compression. To estimate this benefit, on every L2 access, a saturating counter is incremented by the L2 miss penalty if compression can elide a miss and decremented by the decompression latency if the access would have been a hit even without compression. The L2 cache is organized so that up to eight tags can be stored and up to four uncompressed data blocks can be stored. If some data blocks are compressed, the set can accommodate up to eight data blocks. The information in the tag storage (tags for eight blocks and the compressibility of each block) allows us to compute the conditions that increment/decrement the saturating counter.

This technique can thus improve hit rates for applications with large working sets without increasing hit times for applications that do not need high L2 capacity. However, some implementation complexities must also be considered. The larger tag storage increases the L2 area by roughly 7%. A compression and decompression pipeline is required and Alameldeen and Wood show that this can be done in under five cycles. The compression algorithm identifies seven different compressible data classes and attaches a 3-bit prefix to every word in a block. Removing or replacing a line in L2 can be somewhat complex because all data elements in a set are required to be contiguous and sometimes multiple entries may have to be evicted.

Indirect Index Cache with Compression (IIC-C), Hallnor and Reinhardt, HPCA 2005

In their HPCA 2005 paper, Hallnor and Reinhardt [157] first make the argument that compression should be employed at the LLC, memory bus, and in memory. This leads to improvements at each level and reduces the compression and de-compression overhead when going between each level. Second, they build on the insight of Lee et al. [156] to limit compression to the L3 and optimize higher cache levels for low latency. Third, they extend their own prior *Indirect Index Cache (IIC)* design [158] to gracefully handle compressed blocks.

The IIC places a block's tag in one of a few possible locations. The tag carries a pointer to the block's location in the data array. Thus, the data array is fully-associative at the cost of a more expensive tag array. When compression is introduced (the *IIC-C* design), a block is scattered across four sub-blocks; all four sub-blocks are used for an uncompressed line and 0, 1, or 2 sub-blocks are used for compressed blocks. The tag must now store a pointer to each sub-block. Thus, the tag storage becomes even more expensive (it is also required to have twice the entries as the data array). This overhead can be as high as 25% for a 64 byte block, hence, the techniques are only effective for larger sized blocks and sub-blocks. However, the flexibility in data placement removes the replacement complexity introduced by Alameldeen and Wood's design [155].

To further reduce the latency cost of decompression in the L3, Hallnor and Reinhardt augment IIC's generational replacement policy. The data array is organized as prioritized FIFO pools. On a miss, a block must leave its pool; depending on whether it was accessed, it moves to a pool with higher or lower priority. When a block in L3 is touched, it is decompressed. The decompressed version is placed back in the L3. If there are no misses in the pool, the block remains there in uncompressed form, and its accesses can be serviced quickly. If the pool suffers from many misses, the uncompressed block will eventually move to a different pool; at that point, it is compressed again. Thus, applications with small working sets will not suffer much from decompression latency, while applications with large working sets will frequently compress their blocks to create space. This achieves a similar goal to Alameldeen and Wood's adaptive cache compression in a completely different manner, but it does require many block copies and compression/decompression cycles.

Compression and Prefetching, Alameldeen and Wood, HPCA 2007

Some of the topics in this section can have synergistic interactions. We earlier mentioned Qureshi et al.'s evaluation of a combined compression and useful-word technique [154]. We next describe Alameldeen and Wood's evaluation of a combined compression and prefetching technique [116]. In essence, the two techniques are complementary, and the speedup with both techniques is greater than the product of the speedups with each individual technique. This is because prefetching suffers from bandwidth pressure and cache pollution, both of which are alleviated by compression; likewise, the latency cost of decompression can be hidden by prefetches into the L1. A key contribution of Alameldeen and Wood is an adaptive prefetch scheme that throttles prefetches based on the cost/benefit of prefetched blocks. The benefit is estimated by tracking use of prefetched blocks. The cost is estimated by leveraging the existence of extra tags in sets that have not been fully compressed (the cache organization is based on their prior work described earlier [155]). These extra tags keep track of recently evicted blocks; this helps track instances where prefetched blocks evict other useful blocks. The above events are tracked with a saturating counter that disables prefetching when it saturates in one direction. This is especially important in multi-core chips where prefetching is shown to be not as effective.

CHOP, Jiang et al., HPCA 2010

We end this section with a technique that is closest in spirit to being a block-level optimization. In essence, this technique employs very large blocks and chooses to cache a block only if the elements of the block exhibit a high degree of spatial and temporal locality. Jiang et al. [159] consider the use of a large LLC that is organized with embedded DRAM, possibly as a 3D stack. If data is organized in this large LLC at the granularity of typical cache blocks (up to 128 bytes), the overhead for tags would be extremely high. Instead, if data is organized at the granularity of OS pages, many blocks within the page would not be used and memory bandwidth would be wasted. Jiang et al. therefore advocate that the large LLC be used to store data at page granularity and filter mechanisms be designed that only fetch pages that are predicted to be "hot". Hot pages are defined as the most accessed pages that together account for 80% of all accesses and, for the chosen workloads, happen to represent 25% of all pages. A filter cache is an on-chip structure that monitors off-chip accesses to pages that are not yet part of the LLC. By default, a page is not allocated in the off-chip LLC on first touch. Over time, if a page in the filter cache is deemed hot, it is then allocated in the LLC. When a page is evicted from the LLC (based on an LFU replacement policy), it is reinstated in the filter cache with some initial counter value. When a page is evicted out of the filter cache, its counter values can be saved in main memory so that its level of activity need not be later re-computed from scratch. Jiang et al. also employ an adaptive policy that occasionally disables the use of the filtering mechanism if memory bandwidth utilization is low.

3.3   SUMMARY

This chapter first examined different techniques for cache partitioning to maximize throughput and provide QoS. Utility-based way partitioning has been frequently employed and can be implemented with relatively low overhead. Miss rate curves can be constructed with sampled shadow tags that incur overheads of the order of a few kilo-bytes. But ultimately, LLC replacement policies and partitioning appear to be headed in the direction of creative insertion mechanisms. Inserting a block near the end of the priority list for a set (RRIP) is a low-overhead method to retain a block only after it has proved to be useful. Likewise, the insertion point can also be used to prioritize one application over another and achieve implicit cache partitions (PIPP). Insertion point mechanisms need very minor logic and storage because they can evaluate competing policies with a few set samples and counters. In addition to computing an optimal cache partition for a workload, the operating system must also try to create a schedule that allows complementary applications to simultaneously share a cache. Some recent papers have used miss rate curves to construct such schedules (ACCESS, DIO). QoS mechanisms similarly need miss rate curves and various run-time statistics to compute resource allocations and monitor if QoS guarantees are not being met. The difference in the many QoS papers is primarily in the resource (cache, bandwidth, ways, molecules, etc.) that is allocated and how that impacts program IPC.

For replacement policies, novel insertion mechanisms appear to be the way of the future because of their simplicity. While some studies have attempted creative replacement policies that approximate Belady's OPT algorithm, they inevitably incur higher overheads in terms of storage and logic complexity. Similarly, many of the policies that boost associativity incur non-trivial complexity. They may have to fight an uphill battle for commercial adoption, especially since software schemes (page coloring or scheduling) can help balance load across sets.

Most commercial processors dominantly use stride or stream based prefetchers. Some recent works have proposed policies that examine behavior to control the parameters of stride-based prefetch engines. These policies appear simple enough to be adopted commercially. There has also been extensive research on prefetching for streams (data blocks with no apparent pattern in their addresses). While these techniques are effective, they frequently require that the streams be stored in memory, a feature that may encounter scalability challenges. Dead block predictors have made significant advancements, but they have been primarily useful when prioritizing blocks for eviction. The sampling dead block predictor of Khan et al. may be simple enough that it could compete with other insertion-based replacement policies. Some block-level optimizations (compression, filtering) lead to variable-sized blocks. Cache organizations that support variable-sized blocks entail significant complexity for tag and data placement. Eventually, the recent research in caching policies will also have potential at lower levels of the hierarchy, with CHOP being an example of a policy that is applied to an off-chip DRAM cache.

CHAPTER 4

Interconnection Networks within Large Caches

Storage cells within an SRAM cache have undergone relatively minor changes over the decades. The SRAM storage cell continues to be the traditional 6-transistor design, although novel variation-tolerant alternatives are being considered for the future (see Section 5.5).

On the other hand, the interconnects used to connect cache sub-arrays have been the focus of several innovations. It has been well-known for over a decade that on-chip wires do not scale as well as transistor logic when process technologies are shrunk [160]. This is especially true for global long-distance wires. Since cache footprints have not changed much over the years (caches continue to occupy approximately 50% of the chip area), wire lengths within caches are relatively unchanged, but they represent a bigger bottleneck, both in terms of delay and energy (quantified shortly). This chapter describes several strategies to alleviate the wiring bottleneck in large cache hierarchies.

4.1 BASIC LARGE CACHE DESIGN

4.1.1 CACHE ARRAY DESIGN

A simple cache consists of a set of SRAM cells organized as multiple two-dimensional arrays to store data and tags. Figure 4.1 shows the organization of a cache with a single data and tag array. Each memory cell is equipped with access transistors that facilitate read or write operations, and the access transistors are controlled through horizontal wiring called wordlines. A centralized decoder takes the address request and identifies the appropriate wordline in the data and the tag array. To reduce the area overhead of the decoder and wiring complexity, a single wordline is shared by an entire row of cells. For every access, the decoder decodes the input address and activates the appropriate wordline matching the address. The contents of an entire row of cells are placed on bitlines, which are written to the cells for a write operation or delivered to the output for a read operation. Although the data placed on the bitlines can be used to directly drive logic for small arrays, because of large bitline capacitance, the access latency of such a design can be very high even for moderately sized arrays. To reduce access time, sense amplifiers are employed to detect a small variation in bitline signal and amplify them to the logic swing. The sense amplifiers also limit the voltage swing in bitlines and hence their energy consumption.

The energy and delay to read or write an SRAM cell itself is very small; it takes only a fraction of a CPU cycle to read an SRAM cell and dynamic energy consumption in SRAM cells is insignificant. The cache access latency is primarily dominated by the wire delay associated with decoder, wordline, bitline, and output bus. Similarly, the number of bitlines activated during a read or write and the length of the output bus determines the dynamic power of a cache.

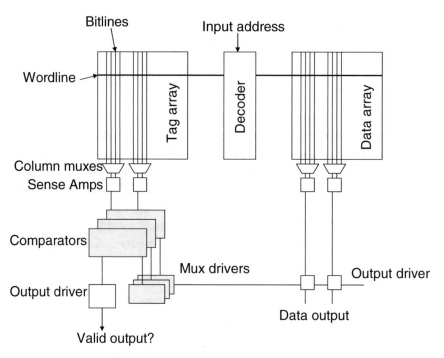

Figure 4.1: Logical organization of a cache.

4.1.2   CACHE INTERCONNECTS

The monolithic model shown in Figure 4.1 does not scale with the array size. Due to low-swing signaling employed in bitlines and silicon area overhead associated with repeaters, wordlines, and bitlines cannot be repeated at regular intervals, causing their delay to increase quadratically with the size of the array. Also, the throughput of a RAM is a function of wordline and bitline delay, and the bandwidth of a single array deteriorates quickly as the array size grows. To reduce this quadratic delay impact, a large array is typically divided into multiple sub-arrays. Many sub-arrays constitute a single bank. All the sub-arrays within a bank share the input address signals and the input/output data signals. A large cache bank is almost never multi-ported. A bank may deal with multiple requests at a time if pipelining is employed. A large L2 or L3 cache is typically partitioned into banks, with each bank operating independently. Each bank can simultaneously receive/service a single request in the same cycle.

An interconnect system is required within a bank to connect sub-arrays. A more complicated interconnect system is also required to connect banks, cores, and other components. The intra-bank system is typically an H-tree network (as shown earlier in Figure 1.7). An H-tree network offers uniform delay to each sub-array, greatly simplifying the necessary scheduling logic. In Section 4.2.2, we discuss the role of the H-tree network and potential optimizations. Most of this chapter will focus on the more problematic inter-bank network.

In relatively small-scale multi-cores, the inter-bank network is composed of a crossbar switch. For example, the Piranha chip (circa 2000) has eight cores and a 1 MB L2 cache organized into eight independent banks [21]. An *intra-chip switch (ICS)*, which is essentially a cross-bar, is used to connect the cache banks to the cores. A cache request must therefore travel a few long wires and go through arbitration for the output port, possible buffering if the request is not immediately granted, and crossbar logic to reach the cache bank. The same is true for Sun's Niagara processor, where a crossbar is used to connect eight cores to four banks of a 3 MB L2 cache [161]. Kumar et al. [162] provide a description and analysis of the interconnects that may be required to connect a medium number of cores and cache banks.

As the number of cores and cache banks scales up, the complexity of the crossbar and necessary wiring can increase dramatically. It is also inefficient to travel half-way across the chip to reach a centralized crossbar. It is generally assumed that tens of cores and cache banks will have to be connected with scalable networks that do not incorporate centralized shared components. In fact, future large caches will be distributed on chip and techniques such as page coloring will ensure that data for a thread will likely be resident in nearby cache banks. Therefore, there must be a way for nearby components to communicate without having to go through centralized structures. Scalable networks can take many forms, but the predominant one adopted in recent literature is the packet-switched routed network [163, 164]. Such a network incorporates many routers, independent links connect routers and components to routers, and packets are sent via multiple routers from source to destination. These networks are more scalable because multiple packets are simultaneously in transit over different links, the centralized functionality of a crossbar is now distributed across several routers, and if cores tend to access data in nearby cache banks, short distances are traveled on average.

4.1.3 PACKET-SWITCHED ROUTED NETWORKS

We'll next provide a short primer on the design of packet-switched routed networks. The interested reader is encouraged to check out the Synthesis Lecture on On-Chip Networks [11] for more details.

Topology   The most fundamental attribute of any on-chip network is its topology that describes the edges required to connect all the routers and components. Common on-chip topologies, in order of increasing number of edges (and hence performance and cost) include the 1-D array, ring, 2-D array or mesh, torus, flattened butterfly, and hypercube. The mesh is by far the most commonly used topology, where components and routers are laid out in a 2-D array and each router is connected to the adjacent component and the four North, South, East, West neighboring routers. Except for the routers on the periphery of the network, each router has a *degree* of five, *i.e.*, each router has five input and five output ports.

Routing Algorithm   The routing algorithm is typically classified as either *Oblivious* or *Adaptive*. An oblivious routing algorithm uses paths that are only a function of the source and destination and not of events or load in the network. A prominent subset of oblivious routing algorithms is the set of *Deterministic* routing algorithms. Deterministic routing employs a unique pre-determined route for every source-destination pair. A common example of deterministic routing is *dimension-order routing* where a message is first routed along the first topology dimension until it reaches the same co-ordinate in that dimension as the destination; the message is then routed along the second dimension, and so on. For example, in a 2-D array, if the source is denoted as (3,5) and the destination as (7,9), the message is first routed along the X dimension until it reaches (7,5), and it then travels along the Y dimension until it reaches (7,9). On the other hand, adaptive routing may employ pre-determined paths as a guideline, but a message may deviate from this path in an attempt to avoid congested routers. A subset of adaptive routing algorithms is the set of *Minimal* routing algorithms. Minimal routing allows the path to be determined dynamically based on load, but it ensures that every hop is in a direction that brings the message closer to its destination.

Buffer, Crossbar, Channel   Every router has a crossbar switch that allows data on any input port to be transferred to any output port. If multiple incoming messages wish to go on the same output port, an arbitration unit must grant the output port to one of the messages. The granting of the output port is equivalent to the granting of the link (also frequently referred to as *channel*) connected to that output port. The losing message is then buffered at its input port. Each input port has a set of buffers. There are two primary physical resources that every message must compete for and acquire in order to make forward progress: one is the physical channel for its next hop and the other is a free buffer entry at the input port of the next router it is visiting. For example, if a message is going from router A to router B and then router C, it must compete to acquire the channel connecting A and B, and it must also confirm that there is a free buffer entry at B's corresponding input port to accommodate this message in case the message is not immediately able to hop to C (because of contention from other messages that may be simultaneously arriving on B's other input ports). The term *flow control* refers to the policy used to reserve resources (buffers and channels) for a message as it propagates through the network.

Packets   Before we discuss flow control options, let us first describe the various granularities that a message can be partitioned into. A message is partitioned into multiple *packets*. Each packet has a header that allows the receiving node to re-construct the original message, even if the packets are received out of order. Since a packet is a self-sufficient unit, each packet can be independently routed on the network and different packets of a message may follow different paths. This can lead to better load distribution in a network with an adaptive routing algorithm. In an on-chip network, messages are typically short, rarely exceeding the size of a cache line. Messages are therefore rarely broken into smaller packets, and the two terms are synonymous. Common packet classes include:

data (64 B cache line), request (4 or 8 B address), control (coherence messages that may be a few bits or up to 10 B if they include the address).

Flits   Packets are further partitioned into *flits*. A flit is typically the same size as the width of the channel and therefore represents the smallest granularity of resource allocation. If the channel is 64 bits wide, and a 64 byte (cache line) packet is being transmitted, the packet is organized into eight flits that are transmitted on the channel over eight cycles. Flits do not carry additional headers. The flits of a packet can not be separated, else the receiver will not know how to re-construct the packet. All flits of a packet therefore follow the same path and are transmitted in exactly sequential order. The head flit carries header information for the packet, and this is used by the router to decide the output port for the packet. Once this decision has been made, all subsequent flits of the packet are sent to the same output port until the tail flit is encountered. With such an architecture, packets can be made as large as possible (to reduce header overhead), but resources can be allocated at the finer granularity of a flit, thereby improving overall network resource utilization. To appreciate this, consider the following analogy: the volume of a bucket is better utilized if we attempt to fill it with sand instead of rocks, *i.e.*, an entity has a better chance at propagating itself through the network if it demands fewer resources. It is worth pointing out that while a channel may be 64 bits wide, it will have additional wires (as much as 20% more wires) for control and ECC signals. Packet headers may also include checksums or ECC codes. Occasionally, if the physical channel width is less than a flit, a single flit may be transmitted across a channel in multiple cycles; the transmission per cycle is referred to as a *phit*.

Flow Control   Flow control policies differ in the granularities at which buffers and channels are allocated. These policies are first categorized as either *bufferless* or *buffered*.

Bufferless Flow Control   Bufferless flow control is potentially power-efficient because it does away with buffers in the network. Studies have attributed about a third of total on-chip network power dissipation to router buffers [165, 166, 167, 168]. In what is referred to as "hot-potato" or "deflection" routing, an incoming flit is sent to another free output port if the desired output port is busy. If the number of input ports matches the number of output ports, it should always be possible to forward all incoming flits to some free output port. This increases the number of hops and latency required for an average message but simplifies the router. Such designs are being strongly considered for modern energy-constrained multi-cores [169]. A *circuit-switched* network is another example of bufferless flow control. When a message must be sent, a short probe message is first sent on the network. This probe finds a path all the way to the destination, reserving all the channels that it passes through. An acknowledgment is sent back to the sending node. The actual message is then sent to the destination, and there is no need for buffering at intermediate routers because all required channels (output ports) have already been reserved. The tail flit of the message de-allocates the channels when it goes through. There is a clear latency overhead with this flow control mechanism because of the round-trip delay of the initial probe message. This overhead can be amortized over large message transfers, but as we've pointed out, messages tend to be short in on-chip traffic. While this technique does not require buffering for the message, it does require some limited buffering for the probe message as it waits for desired channels to be vacated.

Buffered Flow Control   Buffered flow control can happen at a coarse granularity with *packet-buffer* flow control, where channels and buffers are allocated per packet. When a packet wants to move from router A to adjacent router B, it must ensure that B's input port has enough free buffer space to accommodate the entire packet. The channel from A to B is then reserved for the packet. Over the next many cycles, the entire packet is transmitted from A to B without interference from any other packet. Two common examples of this flow control are *Store-and-forward* and *Cut-through*. In the former, the entire packet must be received by B before the header flit can begin its transfer to the next router C. In cut-through, the header flit can continue its traversal to C before waiting for the entire packet to be received by B. Cut-through often leads to lower latency, but it can be more resource-intensive because an in-transit packet is scattered across multiple routers and packet-sized buffers are reserved at each router. *Wormhole flow control* improves upon the storage inefficiency in cut-through by allocating buffers on a per-flit basis. This is a form of *flit-buffer* flow control, but channels continue to be reserved on a per-packet basis, *i.e.*, once a packet begins its traversal over a channel, no other packet can occupy the channel until the first packet's tail flit has been transmitted. With wormhole flow control, a packet may be stalled with its various flits scattered across multiple adjacent routers. When a packet is stalled in this manner, it prevents the involved channels from being used by other packets that may be able to more easily make forward progress. This brings us to the last form of buffered flow control, *Virtual channel flow control*, that also allocates the channel on a per-flit basis.

Virtual Channel Flow Control   In virtual channel flow control, it is assumed that multiple packets can simultaneously be in transit over a link or physical channel. We mentioned earlier that flits do not have headers and the flits of a packet must be transmitted sequentially over a channel. For multiple packets to share a physical channel, each flit must carry a few-bit header that identifies the in-transit packet that it belongs to. When a router receives a flit, the few-bit header is used to steer the flit towards its appropriate packet. Therefore, it is as if multiple virtual channels exist between the two routers and each packet is being transmitted on its own virtual channel and being received by its own set of buffers at the other end. In reality, there may be only one physical channel and one set of shared buffers, but the virtual resources are being multiplexed on to the physical resources. If we assume four virtual channels (VCs), there can be four packets in-transit over a physical channel. Each cycle, a flit belonging to one of the four packets (identified by the 2-bit header) can be transmitted over the physical channel. This time-division multiplexing of multiple virtual channels on a single physical channel allows other packets to make forward progress while some packets may be stalled. At the input port of each router, there needs to be state for each of the four packets. This state must keep track of the output port and virtual channel that has been selected for the next hop of the head flit of each packet; subsequent flits of that packet are directly forwarded to that output port and virtual channel.

The virtual channel flow control mechanism is considered the gold-standard for on-chip networks. It not only yields the highest performance, virtual channels are also used for deadlock avoidance (discussed shortly). For such an architecture, various resources must be acquired before a packet can be forwarded to the next router.

VC Allocation   First, the head flit must acquire a free virtual channel on the link that needs to be traversed. The VC state resides at the input port of the receiving router. The sending router realizes that a VC is available as soon as it sends a tail flit on that VC. Even if the previous packet that used the VC is buffered at the receiving router, the next packet can begin transmission on the same VC; the flits of the next packet simply get buffered behind the earlier packet and will be dealt with after the first packet has been forwarded on. The remaining body flits of the packet need not compete to acquire a virtual channel.

Buffer Management   Second, each flit must make sure that a free buffer entry exists in the next router before it can hop across. This is usually determined with one of two methods. In a *credit-based* scheme, every time a flit is forwarded on by a router, a signal is sent to the preceding (upstream) router to inform it that a buffer entry has been freed. A router is therefore aware of the number of free buffers at all adjacent (downstream) routers, with a slight time delay. In an *On/Off* scheme, the downstream router sends a signal to the upstream router when its buffers are close to being full. The upstream router stops sending flits until the downstream router de-activates the signal. The thresholds for activating and de-activating the signals are a function of the link delay between the two routers. The latter scheme reduces upstream signaling and counters but wastes a little more buffer space.

Physical Channel Allocation   Third, each flit must compete for the physical channel. If we assume that each physical channel has four virtual channels associated with it, up to four flits may be ready for traversal in a cycle, but only one can be transmitted on the physical channel. This is a local decision made at that router.

If we take the example of a 5x5 router (5 input and 5 output ports) with four virtual channels per physical channel, there may be as many as 16 incoming packets (on 4 incoming ports and 4 VCs on each port) that wish to all go on the same output port. The head flits of these 16 packets will compete for the four available VCs on the desired output port. Assuming that free buffers exist in the downstream router, flits from the four winning packets will compete for the physical channel in every cycle.

Basic Router Pipeline   This leads us to the design of a typical router pipeline. There are four essential stages:

- *Routing Computation (RC)*: When a head flit is encountered, its headers are examined to determine the final destination. The routing algorithm is invoked to determine the output port that this packet should be forwarded to. An adaptive algorithm will take various load metrics into account when making this decision. State is maintained for each VC on each input port and the output port is recorded here. Subsequent body flits of that packet need not go through the RC stage; they simply adopt the output port computed by the head flit.

- *VC Allocation (VA)*: Once the output port has been determined, the header flit must acquire a free VC at the selected output port. If a VC is not immediately granted, the head flit must continue to compete every cycle until a VC is allocated. The assigned output port VC is also recorded in the state being maintained for the packet at its input port VC. Again, subsequent body flits skip this stage and adopt the VC acquired by the head flit.

- *Switch Allocation (SA)*: All flits go through this stage. At the heart of every router is a crossbar that allows a flit on any input port to be sent to any output port. A packet's leading flit must compete with other packets for access to the output port (physical channel). The SA stage assigns the available output ports to a sub-set of competing flits.

- *Switch Traversal (ST)*: The flits that are assigned output ports begin their traversal in this stage. The flits are read out of their input port buffers and traverse through the crossbar. This stage is followed by multiple traversal stages on the link itself. For example, if the link has a 3-cycle delay, it is typically pipelined into 3 latched stages.

An example pipeline and packet traversal through the pipeline is shown in Figure 4.2(a). We mentioned that a flit must acquire VC, physical channel, and buffers before a hop – while the VA and SA stages allocate the VC and physical channel, there is no separate stage to allocate the buffer. A quick check of credits or on/off state is done as part of the SA stage.

Speculative Router Pipelines   The router just described has a 4-stage or 4-cycle pipeline. The delay through a router can be reduced with speculative techniques [163, 170]. Example speculative router pipelines are also shown in Figure 4.2. In the first example (b), the VA and SA stages are done in parallel. A flit must assume that it will succeed in acquiring a VC in the VA stage, and it speculatively competes for an output port in the SA stage. The speculation fails only if the flit is a head flit and the VCs are heavily contested. In low-load conditions, such speculation will often be successful and reduce delay through the router. In high-load conditions, speculation may fail, at which point the pipeline is traversed sequentially. The only penalty is that the output port may have been allocated to a flit that did not acquire a VC and hence the output port remains idle for a cycle. In high-load conditions, the bottleneck for a message is the queuing delay in buffers and not the router pipeline delay. It is therefore fortunate that speculation is helpful in low-load conditions where router pipeline delay is indeed the bottleneck.

In the second speculative router example in Figure 4.2(c), the VA, SA, and ST stages can happen in parallel. Flits are directly jumping to the output port, assuming that they will successfully acquire the VC and the output port. This speculation again has high accuracy in low-load conditions. The final example (d) implements a 1-cycle pipeline where all 4 stages are traversed in parallel. While it may be possible to predict the output port selected by the RC stage, this will likely be a low-accuracy prediction. Therefore, we assume that when a flit arrives at a router, it is accompanied by a few bits

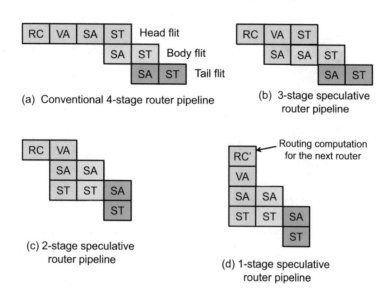

Figure 4.2: Examples of conventional and speculative router pipelines.

that specify the optimal output port for that flit. In other words, the route comes pre-computed. Since the output port and downstream router are known, the RC stage now tries to estimate the output port that should be used by the downstream router. This pre-computed route for the next router is shipped along with the flit to the downstream router. Truly speaking, the router operations are pipelined into 2 sequential cycles but distributed across two routers. This is also referred to as *look-ahead routing*. Assuming that speculation succeeds, the delay at a given router is only a single cycle. We are not aware of commercial routers that employ such a high degree of speculation. It is clear that speculative routers may consume more energy and entail higher complexity.

Deadlock Avoidance   Network routing and flow control algorithms must also incorporate deadlock avoidance features. Deadlocks are created when there is a cycle of resource dependences. Consider a flit that is attempting to hop from router A to router B. The flit is currently holding a buffer entry in router A's input port and it is attempting to acquire a free buffer in router B's input port. The release of buffer resources in router A is contingent on the release of buffer resources in router B – this is a resource dependence. As shown in Figure 4.3(a), four packets attempting a turn can end up in a resource deadlock. Deadlocks are best understood by the four-turn model (Figure 4.3(b)). An unrestricted adaptive routing algorithm can lead to deadlocks because all four possible turns

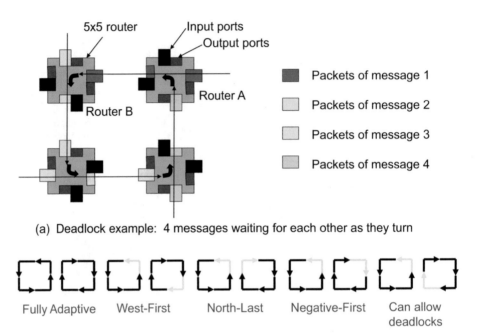

(a) Deadlock example: 4 messages waiting for each other as they turn

Fully Adaptive   West-First   North-Last   Negative-First   Can allow
                                                                deadlocks

(b) The turn model to reason about deadlock possibilities for each routing algorithm

Figure 4.3: Example of deadlock and the 4-turn model.

(in a clock-wise and anti-clockwise loop) are possible, potentially leading to the deadlock shown in Figure 4.3(a). A dimension-order routing algorithm is deadlock-free because it only allows two of the four turns required for a cycle. Virtual channels can also be used for deadlock avoidance. Assume that each VC has a set of buffers associated with it and the buffers are numbered. A packet is allowed to make arbitrary turns as long as it acquires VC buffers in numerically ascending order. This guarantees that a cycle of resource dependences will not happen. When it is clear that higher numbered VC buffers may not be available, the packet resorts to a deadlock-free routing algorithm such as dimension-order routing.

Router Bottlenecks   The reason for delving into router design in such great detail is that on-chip networks are an integral part of access to a large cache. As can be seen by the discussion so far, scalable networks may need routers with complex functionality. While we have focused much of our discussion on a virtual channel router, even simpler routers require similar functionality. Studies have attributed a significant fraction (10-36% [165, 168, 171]) of total chip power to on-chip networks. Therefore, router and network access cannot be abstracted away while considering performance and energy optimizations for large caches.

A router itself has several key components. While energy breakdowns in the literature vary, it is fair to say that the link, buffers, and crossbar contribute substantial portions to a bulk of network energy. The energy and area complexity of a crossbar grows quite significantly with the number of ports and flit width. Buffer energy also grows with the number of ports, VCs, and network load. Arbiters and routing logic consume non-trivial area but do not contribute much to overall energy.

Router Optimizations   Several attempts are being made to address each of these bottlenecks. Many recent router optimizations target crossbar and buffer energy [167, 169, 172, 173]. Link energy is also being reduced with low-swing wiring [174]. Router traversal can be entirely bypassed with the use of express channels. Physical express channels [175] are long-distance links that connect regions of a chip. Packets traveling long distances can hop on physical express channels for the most part and skip traversal through many intermediate routers. The downside is that some routers are more complex because they must accommodate both express and regular channels. The virtual express channel architecture [176] uses a regular topology, but some virtual channels are reserved for long-distance communication. A packet on such a virtual express channel can bypass the entire router pipeline of intermediate nodes. Concentration is another technique that reduces network size and average required hops, but it increases router degree and complexity [177]. The design of an energy-efficient high-performance network and router remains an open problem. This is an actively researched area and not the primary focus of this book. The optimizations listed above are all examples of network optimizations that are somewhat oblivious of the cache architecture, *i.e.*, they will apply to any network and almost any traffic pattern. We will restrict our subsequent discussion of network innovations to those ideas that are strongly related to the organization of the cache hierarchy.

4.2   THE IMPACT OF INTERCONNECT DESIGN ON NUCA AND UCA CACHES

4.2.1   NUCA CACHES

It has been known for a while that monolithic large caches will suffer from long delays and high energy. It is inevitable that large caches will be partitioned into smaller banks, possibly distributed on chip, and connected with a scalable network. Many of the early evaluations of such NUCA architectures made simplifying assumptions for the on-chip scalable network. It was assumed that each network hop consumed a single cycle and network energy was rarely a consideration. This led to early NUCA designs where the cache was partitioned into 64-256 banks, and many network hops were required to reach relatively small 64 KB cache banks. In the past five years or so, there has been much greater awareness of the complexities within a router and its delay and energy overheads. While it may be desirable to partition a cache into several banks to improve delay and energy within cache components, this comes at the cost of increased network delay and energy. This was quantified by Muralimanohar et al. [4, 7, 178] by constructing a combined cache and network model. They show with a comprehensive design space exploration that an optimal large cache organization is achieved when the cache is partitioned into only a few banks. While more energy and delay is dissipated within a cache bank, much less energy and delay is dissipated within the on-chip network.

This is an important guideline in setting up baseline parameters for any large cache study. The design space exploration is modeled within the popular publicly distributed CACTI cache modeling tool [4]. It also provides a breakdown of the energy/delay dissipated within network and cache components, allowing researchers to make Amdahl's Law estimations of the overall benefit of their optimization to a component.

Most of the analysis of Muralimanohar et al. focuses on a physically contiguous large cache. This is representative of the layouts of Beckmann and Wood [2] and Huh et al. [22] where the cache is placed in the center of the chip and surrounded by cores. This may also be representative of a cache hierarchy where a large L2 cache is distributed on chip, and it is further backed by a large physically contiguous L3 cache on either the same die or on a 3D stacked die. Note that the layout of a distributed cache is not generic enough to incorporate reasonably in a model. The CACTI 6.0 model therefore focuses on a physically contiguous large cache but provides the tools necessary to manually model the properties of a distributed cache.

CACTI 6.0 models the usual elements within a cache bank, as described earlier in Section 4.1.1. In addition, it models the routers and wires of the on-chip network. Different types of wires are considered. Traditional RC-based wires with varying width/spacing and repeater size/spacing are considered. Each wire type provides a different trade-off point in terms of delay, energy, and area (bandwidth for a fixed metal area budget). Low-swing differential wires [179] are also modeled; these wires provide much lower energy while incurring delay and area penalties. The router model is based on Orion [166] and primarily quantifies energy for buffers and crossbar. The router delay model is an input parameter as the number of router pipeline stages is a function of the clock speed and degree of speculation. The design space exploration is carried out by iterating over all reasonable partitions of the cache into banks. It is assumed that each bank is associated with a router. For each such partition, the average network and cache delay is estimated per access. The network delay includes router contention, which is empirically estimated for a number of benchmarks and included in the CACTI model. The design space exploration also sweeps through all interconnect options and various VC/buffer router organizations for each partition of the cache. Ultimately, CACTI provides the cache organization that optimizes some user-defined metric.

Figure 4.4 best illustrates the trade-offs considered by the design space exploration. For a fixed size cache, as the number of banks is increased, the bank size reduces and so does the delay within the bank. At the same time, the network size grows and while the overall cache area and wire lengths do not change dramatically, router delays start to play a prominent role. Further, as more links are added in the network, contention reduces. The net result is an overall delay curve that is minimized when the large 32 MB cache is partitioned into eight 4 MB banks. This results in a network size much smaller than that assumed in most prior work on NUCA caches. A similar trade-off curve also exists for overall cache energy.

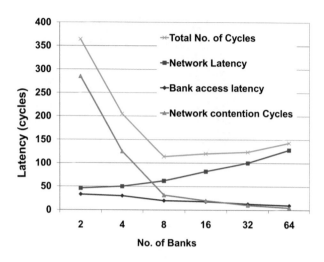

Figure 4.4: Cache access latency as the number of banks is increased [4].

To provide insight into the contributions of various components to delay and energy, we show pie-chart breakdowns for a NUCA hit in Figure 4.5 and Figure 4.6. The results are generated by CACTI 6.5 for a 128 MB L2-cache at 32 nm technology for a 16 core processor. The optimal number of banks for this configuration is found to be 32 banks organized as an 8 × 4 grid. Figure 4.5 shows key components that contribute to NUCA hit latency. There are some interesting observations that can be made from the figure. First, router latency is almost the same as bank access latency in an optimal configuration. Second, contention in the network contributes significantly to the access time. Interestingly, it goes down dramatically (by more than 75%) as accesses to banks are pipelined (not shown in the figure). This shows that having a pipelined bank or multiple sub-banks within a NUCA bank are interesting design points worth exploring. Finally, even with this high contention, the optimal number of banks is only 32. Figure 4.6 shows the energy breakdown. Since each bank is 4 MB, the bank access energy dominates the total energy. Note that not all accesses will have a similar breakdown; if an access is a miss or if a request is for directory information, only the tag-array gets activated, and its access energy is much smaller. In these cases, energy in router components will dominate. If the workloads are mostly single threaded, it is more beneficial to reduce bank energy. On the other hand, for multi-threaded workloads with many coherence transactions, routers and links are good targets for energy optimization. This shows that depending upon the type of workloads being executed, smart voltage or frequency scaling can be done.

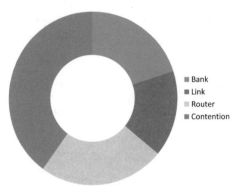

Figure 4.5: Latency breakdown of a NUCA access.

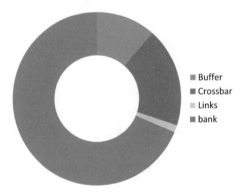

Figure 4.6: Energy breakdown of a NUCA hit.

4.2.2   UCA CACHES

Based on the discussion above, it is clear that a large NUCA cache will be decomposed into one or two handful multi-megabyte banks, not into tens of small banks. This is required to amortize the overheads of bulky routers. A tiled architecture will likely be common where each tile will accommodate a large L2 cache bank (either a private cache or a slice of a large shared cache). Therefore, regardless of whether future processors adopt private/shared caches, or UCA/NUCA architectures, or tiled/contiguous L2 caches, it is evident that several large cache banks will be found on chip. The optimization of such a large cache bank is therefore important. We expect such a bank to have a capacity of a few mega-bytes and offer uniform cache access.

Figure 4.7 shows a pie chart breakdown of where delay and energy is dissipated for various sizes of UCA cache banks, as modeled by CACTI at 32 nm technology. The contribution of the H-tree network to cache bank power clearly stands out as a bottleneck in large cache banks. An

H-tree network is used to connect the many sub-arrays of a cache bank to its input/output ports. A balanced H-tree provides uniform pipelined access without complex switching circuits or scheduling logic. In small cache banks, bitline energy tends to dominate. As the cache bank grows in size, the contribution of the H-tree network begins to dominate. This is primarily because bitlines employ low-swing differential signaling, while the H-tree network uses global RC-based full-swing wiring.

Figure 4.7: Energy breakdown of a UCA access.

The above observations make the argument that low-swing differential signaling [174, 179] also be employed to connect cache sub-arrays. Reducing the voltage swing on global wires can cause a linear reduction in energy. A separate smaller voltage source can lead to quadratic energy reductions. In essence, depending on the length of the wire, low-swing wiring can reduce energy by nearly an order of magnitude. Low-swing wires do not need repeaters, but they do require receiver and transmitter circuits at their end-points. As a result, they cannot be easily pipelined. They also incur longer delay penalties when traveling long distances.

A recent paper by Udipi et al. [180] considers the use of low-swing wiring within a cache bank to connect the many sub-arrays. While dramatic energy reductions are possible, there can also be appreciable performance degradation. Hence, the key is to employ a judicious amount of low-swing wiring to balance the performance penalty and energy saving. The paper puts forth four reasonable options: (i) implementing the H-tree as an unpipelined set of low-swing wires, (ii) implementing the H-tree as a pipelined set of low-swing wires, (iii) implementing multiple parallel sets of low-swing wires, and (iv) augmenting a baseline global-wire H-tree with a low-swing trunk to access two rows of sub-arrays. These different designs fall on different points of the energy-performance trade-off curve. Udipi et al. argue that the last design represents a sweet-spot, especially if architectural mechanisms are used to increase the likelihood of fetching data from sub-arrays with low-swing links. However, these mechanisms involve non-trivial complexity as popular or recently-touched blocks must be migrated to the sub-arrays with low-swing links. Similar to the S-NUCA vs. D-NUCA discussion, there might be alternative less-complex ways to exploit non-uniform power access in cache sub-arrays, but this remains future work. The study does highlight the need to seriously consider the judicious use of some low-swing wiring within a cache bank to target the interconnect bottleneck.

4.3   INNOVATIVE NETWORK ARCHITECTURES FOR LARGE CACHES

It is common for most architecture studies to assume basic mesh topologies when connecting many cores and cache banks. Some studies also employ concentration, express channels, or a flattened butterfly topology. In this section, we will focus on network innovations that are closely tied to their interactions with the cache hierarchy.

Transmission Line Caches, Beckmann and Wood, MICRO'03

A transmission line is a wiring technology that works differently than traditional RC-based wires. In a transmission line, delay is determined by the time taken to propagate and detect a voltage wavefront on the wire. This delay is determined by the LC time constant and velocity of the wavefront, which is a function of the speed of light in the dielectric surrounding the interconnect. A transmission line is therefore capable of very low latencies. However, there are several associated overheads. The wire must have very high width, thickness, horizontal and vertical spacing, high signal frequency, reference planes above and below the metal layer, and shielding power and ground lines adjacent to each transmission line. In addition, transmitter and receiver circuits are required at each end (but there is no need for repeaters). Because of these high costs, transmission lines have only been sparsely used in various chips [181, 182, 183, 184, 185]. Their popularity has not grown in the last half-decade, partially because low latencies can be provided in the RC realm with *fat* wires, and partially because there has been a stronger focus on low-energy interconnects than on low-latency interconnects. That said, they remain an interesting design point worth keeping an eye on. Because of the area/cost overhead, they can likely be exploited only in scenarios where low bandwidth is acceptable.

Beckmann and Wood formulate a large cache architecture, assuming that transmission lines are the interconnect of choice [186]. The L2 cache is partitioned into large banks and placed on the chip periphery. The banks are connected to a central L2 cache controller via transmission line links. These links must be narrow because of the high area overheads of transmission lines. A couple of optimizations are employed to deal with this bandwidth problem. Only part of the address is sent to a bank and a 6-bit partial tag comparison is done at the bank. On a load, the rest of the tag must be returned to the cache controller for full tag comparison. Since the entire address must be exchanged between the banks and controller, there is no saved bandwidth in case of a load. In case of an L1 writeback (store), if a partial tag comparison succeeds, the value is directly written into the L2 – the inclusive nature of the L1-L2 hierarchy guarantees that a partial tag hit corresponds to a full tag hit. The partial tag strategy therefore reduces the bandwidth requirement on L1 writebacks. The second optimization distributes a single cache line across multiple banks. While this increases the overall bandwidth requirement (the address must be sent to multiple banks), narrower links can be employed for each bank as the bank only provides part of the cache line. This is not dissimilar to modern DDRx DRAM data mapping where the limited DRAM chip pin bandwidth is alleviated by distributing a cache line across multiple DRAM chips. The resulting L2 cache yields high performance thanks to the clear low latency advantage of transmission lines.

Halo Network, Jin et al., HPCA'07

Jin et al. [5] show that a network specifically targeted at D-NUCA traffic can be much simpler than a generic mesh-based network. It is assumed that a single core is accessing a large D-NUCA cache. The cache is organized as a 2D array of banks and a given block maps to a unique column, but can reside in any of the banks in that column, *i.e.*, the ways of a set are distributed across all the banks in a column. Placement in the ways of a set is based on the LRU algorithm. So the MRU block is placed in the way and bank that is closest to the core, and the LRU block is placed in the way and bank that is furthest. Assuming that the core is placed above the cache (as shown in Figure 4.8(a)), a cache request is first sent in the X dimension until it reaches the correct column. The request then propagates downward until the block is located in one of the ways (banks). As the request propagates downward, it is accompanied by the block that was resident in the previous bank (way) – this allows the blocks to be placed in banks in LRU order while the cache is being searched. When a block is located in a bank or in the next level of the hierarchy, it is placed in the closest way. When a block is located, it is propagated up the column (in the Y dimension) until it reaches the top-most bank and then propagated in the X dimension until it reaches the core.

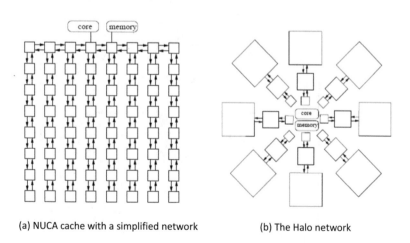

(a) NUCA cache with a simplified network   (b) The Halo network

Figure 4.8: The Halo network [5].

Given this routing of requests and responses, Jin et al. observe that the network is deadlock-free. A full-fledged mesh network is not required as horizontal communication only happens in the top row – all other rows need not have horizontal links. This greatly simplifies the design of each router. The organization is further optimized by re-organizing the banks so they surround the core (Figure 4.8(b)). Requests and responses radiate away or towards the center. Bank sizes can now be heterogeneous, with the smallest banks closest to the core. While the figure shows diagonal links, layouts are possible with zig-zagging XY links as well.

While the design does not extend easily to a multi-core layout and is specially tuned to D-NUCA traffic, it is a thought-provoking idea for cache organizations that may exhibit very specific traffic patterns.

Nahalal, Guz et al., SPAA'08

Guz et al. [6] make the observation that the handling of shared data in a D-NUCA cache is highly sub-optimal. Shared blocks tend to gravitate towards a center-of-gravity where they are not very close to any of the cores [2, 22]. They attempt to solve this problem with a novel layout/topology, that is inspired by urban planning concepts used in the design of a cooperative village Nahalal. While the paper is not specifically about the design of the on-chip network, it shows that clever layout and placement of cores and cache banks can play a strong role in reducing on-chip distances and network traffic.

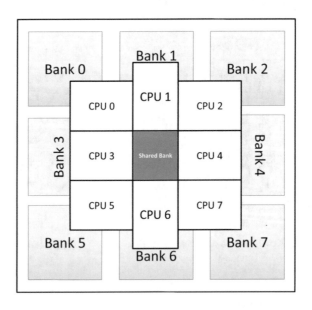

Figure 4.9: The Nahalal architecture [6].

In the proposed organization (shown in Figure 4.9), some cache banks are placed in the center of the chip, they are surrounded by a ring of cores, and another set of cache banks forms an outer ring. The cache banks together form a shared L2 cache, ways are distributed among banks, and blocks can move between ways (banks) with D-NUCA policies. The central banks are meant to store blocks that are shared; the outer banks are meant to store blocks that are private to the nearest core. Blocks are initially placed in the requesting core's outer (private) banks; if a block is accessed by other cores (detectable with a counter or by examining the directory protocol's sharing vector), it is migrated to the central shared banks. Thus, each core is likely to find data in either the shared central banks or the adjacent private banks – this is also the order in which banks are looked up when searching for data. Instead of a parallel look-up in the shared and private banks, a predictor is used to prioritize look-up in one of the two regions. Given the layout of cores and private/shared banks, a large majority of cache look-ups are serviced by traveling relatively short distances. While Guz et al. focus on a D-NUCA architecture, it should be possible to apply these concepts to other S-NUCA or tiled architectures as well.

Hybrid Network, Muralimanohar and Balasubramonian, ISCA'07

As described earlier, wires can be implemented in many ways. Low-latency wires can be designed by having large width and spacing (*fat* wires) or with transmission line technology, but both of these approaches trade off bandwidth because fewer wires can be accommodated in a fixed metal area budget. Muralimanohar and Balasubramonian [7] make the observation that in a large cache network, requests and responses place different demands on the network. Responses require cache line transfers and are in need of high bandwidth. Requests, on the other hand, only consist of the address and possibly even a partial address. They therefore recommend the use of different networks for requests and responses. Responses employ a traditional mesh-based topology with minimum-width wires for high bandwidth. Requests employ low-latency fat wires. Since fat wires can cover relatively long distances in a cycle, a mesh topology may be overkill as it frequently introduces router pipeline delay in data transmission. Therefore, a bus-based broadcast network is advocated for the request network. To improve scalability, a bus is implemented for each row of cache banks and the buses are connected with a regular routed network (Figure 4.10). This work shows that wire-aware design of the cache network and heterogeneity can lead to performance and power improvements.

A similar argument is also made by Manevich et al. [187] in their Bus Enhanced NoC (BENoC) architecture where a multi-stage NoC is augmented with a simplified bus that is more efficient at specific operations, such as searching for a block in a D-NUCA architecture.

Network Innovations for Efficient Directory-based Coherence

Cheng et al. [188] make the observation that different messages in a coherence protocol make different demands of the network. Data messages require high bandwidth, while most other control messages are short and require low latency. In coherence transactions that require multiple messages, each with a potentially different hop count, some messages are not on the critical path – such non-

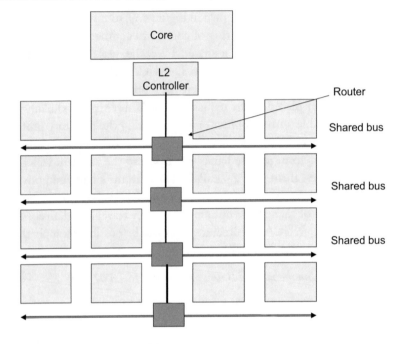

Figure 4.10: The hybrid network for a NUCA architecture [7].

critical messages can be transferred on links that are optimized for energy and not latency. Cheng et al. design a network where each link is made up of different wire types. Each message is mapped to wires that best match its needs in terms of latency, bandwidth, and energy.

Bolotin et al. [189] also make the observation that short control messages in a coherence protocol merit higher priority than long data messages. They also design a broadcast invalidation mechanism that relies on routers to create copies of the invalidation message and quickly spread it to all nodes. This technique can also be used to speed up block search in a D-NUCA cache. Alternatively, given a set of sharers, a virtual ring can be constructed and the invalidation message can be sequentially sent to all the sharers until a single final acknowledgment message is returned to the directory.

Eisley et al. [190] propose a mechanism that adds intelligence within routers to speed up coherence transactions. In a typical directory-based cache coherence protocol, the directory must be contacted before a cache miss can be serviced. The directory, in turn, may have to contact other nodes before the requestor is finally granted access. Even as the initial request is making its way to the directory, it may pass other nodes that have enough information to service the request, but conventional protocols do not leverage this information in any way. Eisley et al. build a distributed directory by constructing virtual trees; all sharers of a block and its directory home node are made to form a tree. The edges of the tree are stored in a table at each involved router. If a request hits a router that happens to be part of the block's virtual tree, the necessary coherence operations are initiated without having to contact the directory. If the request is a read, a nearby sharer provides data and the requestor is added to the virtual tree. If the request is a write, invalidations are sent along the virtual tree and the tree is de-constructed. Thus, the integration of virtual trees within network routers allows intermediate routers to intercept and respond to coherence requests. If the table is full and a tree entry must be evicted, part of the tree will also have to be de-constructed. In addition to the table introduced in each router, the in-network coherence mechanism also entails additional cache look-ups at nodes that are part of the block's virtual tree but do not have a cached copy of the block. Eisley et al. [191] also apply the same philosophy of in-network intelligence to guide placement of an evicted block in a nearby bank that has room (as in Cooperative Caching [48]).

Network Innovations for Efficient Snooping-based Coherence

It is generally assumed that snooping-based coherence will not scale to large numbers of cores. However, a snooping-based protocol has some desireable features: (i) no directory storage overhead, (ii) no need for indirection via a directory when responding to requests, and (iii) a protocol that is simpler and easier to verify. In order to meaningfully exploit these positive features, some of the negative features of the snooping-based protocol will have to be overcome, viz. a centralized ordering point and the need to send every message to every node.

Agarwal et al. [192] attempt to alleviate the first overhead in an HPCA'09 paper. They assume a packet-switched mesh (or any other scalable topology) unordered network and implement a broadcast-based snooping protocol on top of it. Each injected message is assigned a unique ID, and routers are responsible for ensuring that a node receives messages in the correct numerical order. In a MICRO'09 paper, the same authors tackle the second problem [193]. They implement filters within the network that track the presence of an address (or address range) in various parts of the network. Accordingly, the broadcast of the address is stifled in those parts of the network that don't need to see it.

Udipi et al. [8] employ a similar approach but with a focus on bus-based networks. Two additional observations motivate their design: (i) buses are preferable over routed networks because they do away with expensive routers, and (ii) given plentiful metal area budgets in future processors, scalability can be achieved by implementing multiple buses. The on-chip network is composed of a 2-level hierarchy of buses (Figure 4.11). A core first broadcasts its request on its local bus. A Bloom filter associated with the local bus determines if the address has been accessed beyond the local bus and if the broadcast needs to be propagated. For applications with locality, many broadcasts will not be seen beyond the local bus. If a broadcast must be propagated, arbitration for a central bus is first done, and the broadcast then happens on the central bus and on other local buses immediately after. Additional filters at the other local buses may stifle the broadcast on other local buses. The initial local bus broadcast is not deemed complete until the central bus broadcast (if necessary) is complete – the central bus broadcast acts as a serialization point to preserve coherence and sequential consistency. Single-entry buffers are required at each local bus to accommodate a request that is

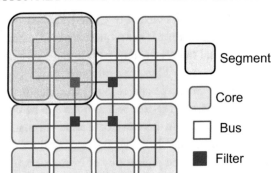

Figure 4.11: A hierarchical bus design with broadcast filters [8].

awaiting arbitration for the central bus. Multiple parallel bus hierarchies can be employed, where each hierarchy deals with a different address range. The energy efficiency of the proposed design, relative to a baseline packet-switched mesh network, is a function of the ratio of link to router energy in the baseline. The proposed design is especially compelling when energy-efficient low-swing links are employed.

Some prior work by Das et al. [194] also argues for the use of hierarchy within on-chip networks to exploit application locality (especially when page coloring is used for smart data placement in an S-NUCA cache). Buses are used for local communication and a packet-switched mesh topology is used to connect the many buses. The use of a local bus helps eliminate router navigation for nearby communication. The use of a directory protocol and the mesh topology at the second level of the hierarchy ensure that messages travel beyond the local bus only if remote sharers exist.

Summary

There are clearly many aspects of the on-chip network that can be improved, such as wiring, topology, router design, etc. Most of the discussed papers can be roughly categorized as follows:

- *Better wiring:* transmission lines [186], fat wires [7, 188], low-swing wires [8].

- *Better topologies:* Halo [5], Nahalal [6], hybrid/hierarchical topologies [7, 8, 187, 194].

- *More/less router functionality:* optimizing for coherence operations [189, 190, 191, 192, 193, 194].

Note that each of these innovations has a specific connection to the design of the cache hierarchy or coherence protocol. There is of course another large body of work that attempts to improve the efficiency of generic on-chip networks [11].

CHAPTER 5

Technology

Chapters 2 and 3 largely focused on logical policies for data selection and placement. Chapter 4 examined the physical design of caches, with a special emphasis on the network; this is an example of how technology trends have significantly impacted the cache hierarchy. Technology continues to improve, presenting new opportunities and challenges for cache design. It is therefore important to understand some of the emerging technology phenomena that may impact cache design. It is also important to understand new emerging memory technologies because the off-chip memory hierarchy may grow deeper and well-known on-chip caching principles will have to be adapted for other levels of the memory hierarchy. On-chip cache policies will also have to be improved to work around the problems posed by off-chip memories.

This chapter begins with a discussion on basic SRAM cells. The basic cache SRAM cell has undergone little change for a few decades. However, advancements in process technology are creating significant challenges and opportunities for cache designers. We provide an overview of these technology advancements and the corresponding cache innovations that have emerged. We end with a discussion of emerging memory technologies.

5.1 STATIC-RAM LIMITATIONS

For more than two decades, static-RAM (SRAM) technology has been an ideal choice for on-chip storage. This is not only because of its superior latency and scalability properties, it is also because the same CMOS transistors used in logic components can also be used in SRAM caches. An SRAM cell is a simple bistable latch that operates at the same voltage as other components in a processor. Hence, it does not require a separate voltage plane or charge pumps commonly used for other memory types. Since the cross-coupled inverters in a latch can automatically correct for any charge leakage in the value stored, the device can hold its value without periodic refreshing, as required in dynamic-RAM (DRAM) memories. By design, it outputs a differential signal, which is easier to detect and more reliable. These characteristics made SRAM the best choice for on-chip storage for many years.

However, as we move to the multicore era, SRAM faces two critical challenges. First, diminishing feature sizes make it increasingly difficult to control transistor dimensions precisely, which increases the leakage power and delay variation in transistors. Further, SRAM is also vulnerable to radiation induced soft-errors, which can lead to random bit flips. High energy particles such as neutrons and alpha particles from space can strike memory cells, disturb their charge balance, and can flip the bit stored in them. With large transistors, the probability of a bit-flip is low and simple error correcting codes (ECC) are sufficient to identify and recover from a few failed bits.

However, soft errors are expected to grow exponentially as transistors get smaller [195]. As a result, more complex and expensive error correcting schemes are required to guarantee correct operation of future SRAM caches. Second, future caches are limited by the high area requirement of SRAM. An SRAM cell is made of six transistors and occupies 20 times more area than a DRAM cell with a simple capacitor and an access transistor. As processors get faster compared to main memory, it is necessary to reduce cache miss rates by having larger caches. The high area cost of an SRAM cell is, therefore, a serious concern in multi-core processors as large caches reduce the real estate for cores for a given die size. This tight area requirement also makes it difficult to adopt alternate SRAM implementations, which trade off area to improve reliability, especially for large caches. For instance, an eight transistor SRAM design separates read and write port within a cell to improve static noise margin [196]. An SRAM cell with ten transistors further increases noise margin by using the extra transistors to provide a strong negative feedback against bit flips [197]. However, their application will mostly be limited to small scratch ram or L1 caches.

Long term solutions to these problems can either try to modify the microarchitecture of caches to make them more robust to failures, or leverage emerging manufacturing techniques (such as 3D stacking) to improve density and resiliency in caches. Efforts are also underway to find new storage technologies suitable for future processors with hundreds of cores. This chapter describes techniques that have been considered for each of these SRAM challenges.

5.2 PARAMETER VARIATION

When fabricating transistors with nanometer dimensions, the ability to accurately reproduce finer lithographic patterns on die reduces. Photolithography is a critical step in the fabrication process, in which circuit patterns are transfered to the silicon die. While Moore's law continues to scale transistor sizes, the wavelength of light used during the photolithography process has difficulty scaling beyond 193 nm. The use of 193 nm wavelength light to produce sub-wavelength patterns ($< 90\,nm$) leads to imperfections in transistor/wire dimensions and threshold voltage (V_t). More specifically, L_{eff}, the effective gate length of a transistor, which has the smallest dimension, is impacted most. Another cause of process variation is the inability to precisely control the doping concentration of active regions in transistors. This again changes the threshold voltage (V_t) of transistors.

These variations severely impact transistor/wire delay and leakage power. Equation 5.1 shows the relationship between gate length, V_t, and delay. As transistor delay is a function of both V_t and L_{eff}, and changes non-linearly with V_t, the effect is more pronounced in transistors than wires. The leakage power varies exponentially with V_t as shown in equation 5.2. Borkar et al. [198] showed that process variation can degrade operating frequency of a processor by more than 30% and increase leakage power by $20X$.

$$tr_{delay} \propto \frac{V_{dd} L_{eff}}{\mu (V_{dd} - V_t)^{\alpha}} \qquad (5.1)$$

$$P_{leakage} \propto V_{dd} T^2 e_t^{-qV} / kT \qquad (5.2)$$

A simple method to control leakage and delay of transistors is to modify the substrate voltage to offset V_{th} changes, referred to as body-biasing. However, body biasing incurs a large area overhead. As variation is highly unpredictable, it is difficult to provision enough biasing circuits at a fine granularity to accurately fix all affected cells and circuits. Hence, architectural solutions are necessary to mitigate process variation without significantly compromising cost or performance. Existing cache microarchitectural solutions for this problem fall in two different categories: isolating inferior cache lines to avoid or limit their usage, or tolerating the extra delay in affected cells.

The effect of process variation, usually, is more pronounced in caches than other logic components in a processor. Due to large die area dedicated for caches, their leakage contribution to the total power is already high. This gets significantly worse because of variation. Furthermore, caches have many critical paths with shallow logic depths, making them more susceptible to the delay variation [199]. As all bits in a word are accessed in parallel, variation in accessing a single bit will slow down the entire read or write operation.

Before describing techniques to tolerate variation, we will briefly look at the methodology typically adopted to model process variation in caches.

5.2.1   MODELING METHODOLOGY

In the past, variation was a concern only between different dies in a wafer. The operating frequency and power characteristics of dies varies across a wafer with some dies having better latency or energy properties than others. However, beyond 90 nm, variation within a die has emerged as a major problem. As a result, even within a single processor, some components are significantly slower or consume more power than others.

Before studying architectural techniques to alleviate the effect of slow components, it is necessary to get an estimation of the deviation in power/latency between components and calculate what percentage of a processor is affected. Process variation can be both random and correlated. To evaluate them, the following five parameters should be considered: transistor gate length, threshold voltage, metal width, metal thickness, and dielectric thickness between layers. For every process generation, ITRS [200] projects targeted values for each parameter. These values are then adjusted based on the variation distribution, variation percentage, and distance or range factor. Depending upon the variation distribution, the variation percentage can be used to model random variation, and the range factor helps model the correlation of variation between nearby components.

5.2.2   MITIGATING THE EFFECTS OF PROCESS VARIATION

As discussed earlier, the two major side-effects of process variation are the increase in device and interconnect delay and increase in leakage power. Since the access latency of a cache bank is often determined by the slowest component, a few slow transistors and wires can have non-trivial impact on the access time.

Although all SRAM caches are affected by variation, their impact on the processor's power and performance depends on the cache's position in the hierarchy. Small high-level caches such as L1-data and L1-instruction have a higher impact on performance than power, while large low-level caches mostly contribute to high leakage power. We next discuss architectural techniques proposed to tolerate variation in both high- and low-level caches.

Yield Aware Caches: Ozdemir et al., MICRO 2006

Ozdemir et al. were among the first to study variation in L1 caches, and they explored microarchitectural solutions to mitigate its effect [201]. Through detailed SPICE simulations, Ozdemir et al. show that inferior transistors and wires in a 16 *KB* cache can degrade the cache access latency by as much as 3X and increase the leakage power by 8X. These poor characteristics can ultimately affect the overall manufacturing yield as microprocessors, even when operational, get tossed away if they fail to meet latency and power constraints.

Ozdemir et al. proposed three schemes to improve the processor yield: Yield Aware Power Down (YAPD), Horizontal-YAPD, and Variable Latency Cache Architecture (VLCA). In all three schemes, the affected components in a cache are first isolated at way or array granularity. Later, based on how severely these components are affected, they are either turned off permanently to achieve the best possible access time and leakage or operated at a low frequency. YAPD and H-YAPD trade off cache capacity to retain fast access latency by turning off ways while VLCA retains the capacity by modifying the out-of-order core architecture to tolerate the poorly performing ways.

YAPD is a simple solution and functions similarly to Albonesi's Selective Cache Ways [202]. If the overall power consumption of a cache or its access latency exceeds a pre-determined threshold, the way with the highest leakage power or the slowest critical path is turned off. The primary problem with this scheme, however, is the granularity at which the optimization is performed. Since a bitline is shared by multiple sets, turning off affected cells, their associated bitlines, precharge circuitry, and sense-amplifiers requires reducing the effective number of ways in many sets; in small caches, all sets will get affected. Also, multiple ways in a set are typically placed close to each other, and their variations are strongly correlated. Hence, the likelihood of two ways in a set performing poorly is very high; in which case, this technique can significantly reduce the effective cache size.

A simple extension to YAPD, called H-YAPD, addresses these problems by turning off horizontal rows in a sub-array instead of a full column. Since a row of cells corresponds to a single set, implementing H-YAPD in a traditional cache will turn off an entire set. To avoid this, cache blocks are organized such that each row in an array is made to store ways from multiple sets. In this way, even if an entire row gets turned-off, the capacity impact on a single set is reduced.

Unlike the above two proposals, VLCA retains poorly performing cache ways and supports non-uniform access to different ways based on their variation factor. The characteristics of VLCA are very similar to NUCA architectures discussed in Section 1.3. However, the novelty of this work lies in extending the variable latency design to L1 caches. For instance, a typical out-of-order core dispatches instructions even before their operands are ready such that the operands get forwarded directly to the execution unit just in time for the execution phase. To support a variable latency load, its dependent instructions dispatched without operands need to be delayed until the load value is ready. VLCA adds load-bypass buffers to all functional units, which allow instructions to wait in the functional unit until a delayed load arrives. While VLCA does not compromise cache capacity, supporting variable latency loads in an out-of-order core has a non-trivial impact on complexity and performance.

Ozdemir et al. also attempt a combination of VLCA and H-YAPD to maximize yield. However, their results show a significant improvement in yield is possible with the simple YAPD design. Yield losses can be reduced by as much as 68% with YAPD, and the most complex hybrid design with both VLCA and H-YAPD improves this to 81%.

Way Prioritization, Meng and Joseph, ISLPED 2006

Meng and Joseph proposed *way prioritization* to tolerate variation in large lower level caches such as L2 or L3. Unlike L1, access latencies of L2 or L3 are less of an issue. However, leakage is a critical problem due to their large size. Similar to YAPD discussed earlier, way prioritization also relies on selective cache ways proposed by Albonesi [202] to turn off slow and leaky cells. However, instead of turning off inferior cache ways permanently, they dynamically change the cache size based on workload requirements. For instance, if a program achieves significant benefit from a large cache, then the leaky ways are employed, provided the decrease in overall processor energy is significant enough to compensate for the overhead of leaky cells. This is done by maintaining a priority register for each bank that tracks the leakage power for each way. At runtime, depending upon the workload requirement, ways are turned on/off so that the overall energy delay product of the processor is optimal.

Substitute Cache, Das et al., MICRO 2008

The techniques, discussed so far, require modification to the cache microarchitecture to tolerate variation and increase the complexity of the design. Das et al. [203] show that for small caches, it is easier to guarantee uniform fast access time through redundancy. It is a common practice to equip storage arrays with redundant rows and columns of memory cells so that in the event of a hard error, the affected row/column is remapped to a redundant row/column. However, this technique is not feasible when addressing variation since the number of affected cells is much greater than the total redundant rows or columns. As a few failed bits will trigger remapping an entire row or column, it is wasteful to increase redundancy to fix all affected bits.

To address this, Das et al. propose adding a separate structure called *Substitute Cache (SC)*. Unlike proposals discussed earlier, the main data array of the cache is left untouched. However, each cache way is augmented with a fully associative cache with a few data entries and their corresponding indices. The data entry is smaller than a cache line (64 b), and the index corresponds to the subset of the address to identify the word. After the manufacturing process, a Built-In-Self-Test is performed to identify critical words whose access time exceeds the targeted delay. The indices of these words are stored in a non-volatile region and populated in the SC at boot-up. Every time the cache is accessed, its SC is also searched. If there is a hit in the SC, data from the SC is selected over the main data array. Since a substitute cache is much smaller than the main data array, its access time is guaranteed to be smaller than the data array.

While the SC is effective in dealing with delay variation, it doesn't address the second major impact of variation, namely leakage power. Further, SC continues to rely on the original tag array to determine hit or miss status, meaning variation in tag array can still break the delay threshold. For instance, while accessing small caches like L1, both tag and data arrays are accessed in parallel. Depending upon the outcome of the tag access, the output from the data array is either sent out or dropped. Hence, the net access time of the cache is determined by either tag or data array (depending upon which one is greater). For most caches, the data array access time dominates due to its large size. In some cases, especially for small set-associative caches, tag-array access time can be greater. In such cases, it is necessary to further extend SC to include full tags so that high latency tags can also be accessed from SC.

5.3   TOLERATING HARD AND SOFT ERRORS

In addition to variation, failures due to soft errors (due to particle strike) and hard errors (due to V_{th} degradation, electron migration, etc.) are becoming more prominent in future technologies. Since any MOS transistor is vulnerable to these errors, they will likely exist irrespective of the storage technology or advancements in the fabrication process. For instance, DRAM and its variants are also susceptible to radiation induced soft errors. Further, a significant real estate of a cache will continue to use CMOS logic for peripheral circuits, which can fail due to both hard and soft errors. This section describes generic techniques aimed at improving the robustness of a cache irrespective of the source of errors or storage technology employed.

The most popular fault tolerant technique for memories is the use of Error Correcting Codes (ECC) to detect and correct errors. ECC adds redundancy to data bits, and its complexity can vary significantly depending upon the error coverage. It is common practice to employ single error correction, double error detection (SECDED) for every 64 bit word. This requires 8 bits to store an error correcting code - a storage overhead of 12.5%. However, as we try to detect/correct more errors, either the storage overhead or the complexity of the recovery process goes up significantly. The BCH code by Bose and Ray-Chaudhuri [204] can correct up to six errors and detect seven. While its storage overhead is similar to SECDEC, the recovery process is extremely complex. Employing stronger ECC such as 8 bit correction and 9 bit detection (OECNED) incurs a storage overhead of more than 89% - a premium not feasible for area constrained on-chip caches.

A general problem with ECC is that irrespective of how the error rate of a cache changes over time, a constant storage overhead has to be paid upfront at design time to tolerate the worst case failure rate. However, the probability of multi-bit failure is extremely rare, and designing for the worst case scenario is both wasteful and inefficient. The following work by Kim et al. [205] and

Yoon and Erez [206] leverage the fact that a majority of reads or writes are error free and propose techniques that decouple error detection and correction to tolerate more failures at a low cost.

Two-Dimensional Error Coding, Kim et al., MICRO 2007

In a traditional array organization, ECC codes are stored along with a cache line in the same row. A read to a cache line will also read the associated ECC, and similarly when updating a cache line, both data and ECC get updated. Kim et al. [205] argue that this single row (or one dimensional) ECC gets significantly more expensive as the coverage goes up. Instead, they propose a two-dimensional coding scheme using simple parity. The key is to have parity for every few bits in both rows as well as columns. Unlike ECC, parity can only detect errors. However, when a row parity detects an error, all the columns containing the affected cells are checked with column parity to identify the failed cell. The failed bit can be corrected by simply inverting it. This technique is also efficient in handling multi-bit errors. While it is expensive to recover from a failure (since to correct a failed bit, the entire column has to be read, which requires reading every row and picking the appropriate bit), during error free operation, the delay overhead for reads is almost zero. But, since every bit in the cache is tied to both horizontal and vertical parity, a write has to update both vertical and horizontal parity. In traditional 1-D protection, both ECC and data get updated at the same time. However, in a 2-D scheme, since a write updates only one bit in each column, updating a column parity requires reading the old contents to calculate the new column parity. Hence, every write has to be preceded by a read to find the new parity. In spite of these complexities, it is a promising step towards building more robust caches.

Yoon et al., ISCA 2009

Yoon et al. [206] propose decoupling the data and the parity information to reduce the storage overhead of complex codes. Their motivation is also based on the fact that errors are extremely rare when compared to the number of fault free operations. Hence, instead of paying a large storage overhead in on-chip caches for high overhead codes, they propose mapping ECC codes of cache lines directly to the main memory. They explore a two-tier protection scheme, where every cache line is protected by parity or other low overhead code implemented similar to a traditional cache. In addition, a region in DRAM is allocated to store ECC. Whenever tier-1 protection throws an error, ECC is explicitly read from the memory to correct the errors. In this way, it is not necessary to store all ECC information on chip. Since the code size is not limited by on-chip storage, depending upon the type of error rate, more complex codes can be used for tier-2. Hence, the approach scales better to meet rising error rates. However, writes continue to be problematic as updates may be required for multiple tiers.

5.4   LEVERAGING 3D STACKING TO RESOLVE SRAM PROBLEMS

Many recent technological advances have been made in the area of 3D die-stacking. Many architectural innovations have been proposed to leverage this technology [207] to address the high area overhead of SRAM. It is becoming evident that multiple dies in a 3D stack will most likely be used for cache or memory storage, potentially with disparate technologies [9, 208, 209, 210, 211, 212, 213, 214, 215, 216, 217]. In this sub-section, we touch upon a sampling of the 3D literature that is most closely tied to caching policies and network design.

Li et al., ISCA 2006

Li et al. [209] were among the first to articulate the detailed design of the network and cache in a multi-core chip (their work was preceded by a few other related efforts [208, 210, 212, 218]). Instead of implementing the network with a 3D array of routers, Li et al. propose the use of 2D meshes on each die. These planar networks are then linked with inter-die connections that are implemented as buses. Since inter-die distances are very short and a stack is expected to consist of fewer than 10 dies, a bus broadcast is expected to be fast. It is clearly overkill to require that routers be navigated during the inter-die transmission. These buses (or 3D *pillars*) are implemented with inter-die vias that are expected to support reasonably high bandwidths in future technologies. While most routers in a planar mesh are 5x5 routers, some routers must also support traversal to/from the pillar and are implemented as 6x6 routers. Cores are scattered across all dies such that no two cores are stacked in the same vertical plane. This helps reduce thermal emergencies. Thermal emergencies can be further reduced if all cores are placed on the die closest to the heat sink. Cache banks are also scattered across all dies and a D-NUCA architecture is assumed. Cache blocks are not migrated between dies because inter-die latencies are not problematic. Li et al. draw the conclusion that the performance benefit of a 3D topology is better than the benefit of block migration in a 2D NUCA architecture. In other words, the performance boost in moving from 2D S-NUCA to 2D D-NUCA is less than the performance boost in moving from 2D S-NUCA to 3D S-NUCA. Performance is also shown to be sensitive to the number of vertical pillars in the network.

Madan et al., HPCA 2009

In more recent work, Madan et al. [9] consider a heterogeneous stack of dies. A bottom die (closest to the heat sink) implements all the cores and L1 caches. A second die implements large SRAM cache banks. A third die implements large DRAM cache banks. Each cache bank has a similar area as a single tile (core plus L1 caches) on the bottom die. An on-chip network is implemented on the second (SRAM) die; there are no horizontal links between tiles/banks on the bottom and top dies. Each tile is assumed to have an inter-die via pillar that connects the core to the SRAM and DRAM cache banks directly above. The SRAM cache is implemented as an S-NUCA L2 cache. When a tile experiences an L1 cache miss, the request is sent to the SRAM bank directly above on the inter-die vias. From here, the request is sent via the on-chip network to the unique bank that caches that address. Madan et al. implement an oracular page coloring scheme, similar to that of S-NUCA page coloring schemes discussed in Section 2.1.3, that places all private pages in the SRAM bank directly above the requesting tile. Shared pages are placed in the central SRAM banks. Such a policy ensures that a single vertical hop is often enough to service an L2 request. Traffic on the horizontal on-chip network is limited to requests for shared data, and this traffic tends to radiate in/out of the chip center. As shown in Figure 5.1, a tree topology is an excellent fit for such

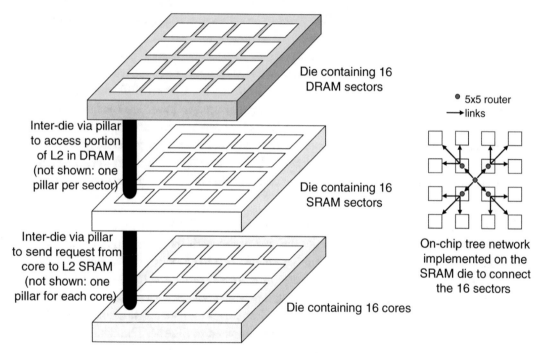

Figure 5.1: A 3D reconfigurable cache with SRAM and DRAM dies and a tree network on the SRAM die [9].

a traffic pattern and helps reduce router complexity in the network. Finally, Madan et al. make the observation that the page coloring policy leads to imbalance in pressure in cache banks and propose a reconfiguration policy. If an SRAM cache bank is pressured, the DRAM bank above is activated to selectively grow the size (associativity) of the L2 cache bank. Part of the SRAM cache bank is now used to implement tags for data stored in the DRAM bank. Thus, there are many options worth considering while organizing a cache hierarchy across multiple (possibly heterogeneous) dies in a 3D stack.

5.5 EMERGING TECHNOLOGIES

Although 3D can boost cache capacity, it does not truly increase the bit density in a silicon die. 3D is also limited by the cost of silicon, power and temperature budget of the chip, and TSV bandwidth. Finally, even with 3D processors, the extra real estate provided is best utilized by balancing cores and caches than completely allocating them for caches. When the cost of additional manufacturing steps required for 3D is weighed in, alternate memory technologies such as DRAM or non-volatile memories (NVM) that are often discounted for caches due to the fabrication complexity can appear compelling. The exponential scaling that can be achieved through a better storage technology likely has higher potential in the long term.

The quest to find a suitable storage technology for multi-cores has gained significant attention in the past few years. This is partly because of the challenges associated with SRAM, and also because other established technologies such as NAND/NOR FLASH have difficulty scaling beyond current process technologies. Though FLASH is primarily used in lower levels of the memory hierarchy, while finding a scalable alternative to FLASH, researchers are attempting to devise a universal memory that can be employed across multiple layers of the memory hierarchy.

When exploring alternatives to SRAM, the following important factors should be considered to minimize performance loss, power increase, or other deleterious effects of the transition.

Area

A cell with the smallest dimension will require an area of $4F^2$ where, F is the feature size. Here, the area occupied by a cell is literally the area underneath the intersection of two minimum sized wires. In comparison, an SRAM cell has an area of $120F^2$. With the advent of *multi-level cells*, it is further possible to increase the density (or reduce effective area/bit below $4F^2$) by storing more than one bit in a single cell. While finding a new technology with cell area close to $4F^2$ or smaller can be challenging, it is not difficult to find alternatives that are much smaller than SRAM. Almost all dominant storage technologies are denser compared to SRAM. The challenge lies in organizing them efficiently to get the maximum benefit out of them.

Access Time

Short access times are essential for caches. Though processor bandwidth is not scaling well, off-chip wires can operate close to the processor frequency [219]. Hence, it is critical to keep the cache access latency much faster than the main memory to justify its presence. An SRAM cell has the best delay characteristics compared to any other storage technology and has uniform read and write latencies. There is also no need for periodic refreshing. But, due to its relatively large size, the wire delay in an SRAM array dominates its access time. Hence, when building caches with alternate technologies, it is necessary to leverage their area advantage to keep the access time low.

The difference between read and write latencies is another major concern in many emerging technologies. For instance, in phase change memory (PCM) and magnetoresistive-RAM (MRAM) – two leading alternatives for SRAM – writes take an order of magnitude more time than reads. This extra delay may not be a problem in uniprocessors where writes are often considered non-critical.

But in multi-cores, the impact of writes on shared data and memory model implementations makes writes more critical. Further, it is necessary to ensure that long writes do not block critical reads.

Other technologies based on DRAM can also block reads due to periodic refreshing required to retain the value. Refresh overhead is typically low, but careful scheduling of refresh operations is necessary to avoid performance penalties.

Endurance

The number of writes to a cache over its lifetime is several orders of magnitude higher than memory or storage disks, and SRAM has the best endurance properties. For instance, an SRAM cell can function even after 10^17 writes while most NVM cells will fail before 10^8 writes. Hence, when adopting NVM with limited endurance for caches, special mechanisms are necessary to keep the cells alive longer. This either requires tuning cells to increase endurance (at the cost of other critical parameters such as energy, delay, density, etc.) or modify the architecture of caches to reduce writes to NVM.

Byte Addressability

The access granularity of a cache is much smaller than the main memory or solid-state storage disks. Hence, technologies that require block accesses such as NAND FLASH can take a big hit in performance when directly employed for on-chip caches. Reading a bulk of data and selecting a few words from it is inefficient and requires high energy.

Candidate Technologies

A few promising technologies pursued for future systems include *Spin Transfer Torque Magnetoresistive Random Access Memory* (STT-MRAM), Memristor, and *Phase-Change Memory* (PCM). Each of them is accompanied with different trade-offs and has to cross significant hurdles to augment or replace SRAM. As a near term goal, another established technology - DRAM and its variants such as *embedded DRAM* (eDRAM) and *Three Transistor DRAM* (3T1D) are being considered for on-chip caches. In fact, eDRAM has already proven commercially viable and is being used in Power 7 processors from IBM [220]. This section presents the landscape of storage technologies for caches and the existing body of work related to them.

5.5.1   3T1D RAM

Besides SRAM, the most widely used and understood technology is DRAM. A DRAM cell consists of a simple capacitor and an access transistor to charge or discharge the capacitor. Due to its minimalistic design, it is significantly denser compared to SRAM. However, the slow charging and discharging operations limits its application to a low level in the memory hierarchy. Latency critical L1 requires a much faster alternative. To address this problem 3-transistor 1-diode (3T1D) cell was proposed, a technology similar to other capacitive memories. It relies on varying charge content of a cell to store logical 1 or 0 [221]. Since the area overhead of 3T1D is relatively higher than other emerging alternatives, it is primarily targeted for L1 caches. There are three key advantages of 3T1D. First, by design, the number of leakage paths in 3T1D is lower compared to SRAM even when process variation is taken into account. Second, instead of a capacitor directly charging or discharging a long bitline, the judicious use of diode as a voltage controlled capacitor to drive a pull-down transistor, which in turn drives the bitline, results in access speed similar to SRAM. Finally, though 3T1D cells require periodic refreshing, unlike DRAM, the reads are not destructive.

In general, the refresh overhead of 3T1D cells have negligible impact on performance. However, Liang et al. [222] show that under severe process variation, some cells require more frequent refreshing than others, and adopting the worst case refresh period for the entire cache can significantly degrade performance and increase refresh power. To address this problem, they propose line-level refresh policies along with novel replacement policies specific to 3T1D caches. The key idea is to augment each line with a hardware counter to track retention times so that lines with high refresh overhead can be served separately. Similarly, while storing a new cache line, the retention of various ways are also considered to find the optimal replacement block.

The paper then studies three different refresh policies with varying complexities. In the simple scheme, a line is either evicted or invalidated as soon as its retention time falls below a certain threshold. As an improvement to this, in a second scheme, refreshing is done but limited to those lines whose refresh interval is below the threshold value. Hence, only cells affected by process variation get refreshed; the remaining lines are evicted when it is time for refresh. The third scheme aggressively refreshes all lines based on their retention value. Interestingly, their study showed that lifetime of a majority of cache lines in L1 is much smaller than the typical refresh interval. Coupled with 3T1D's support for non-destructive reads makes it a promising alternative for SRAM in L1 caches. The second scheme (partial refresh) achieved even better performance consuming lesser power; although, the performance difference between different policies is less than 2% even with severe process variation.

Another factor that can impact refresh overhead is the replacement policy. Typical L1 or L2 caches adopt *LRU* policy to evict a cache line. However, LRU is not well suited for caches with some blocks that have poor retention time, and some that are completely dead (blocks with retention time less than a certain threshold). For instance, a dead block will get invalidated soon after storing a new block and will likely show up as the least recently used block more frequently. To address this issue, the paper studied different replacement policies. In their most simple scheme, traditional LRU policy is extended to avoid using dead blocks while placing a new cache line. Second, in addition to avoiding dead blocks, a new line is always placed in a slot with maximum retention time. The replaced block is moved to the next slot within the set with the longest retention, and the process is repeated until the block with least retention time is evicted. The third scheme is similar to the second except that the most recently used block is always placed in the best slot. A detailed performance study of these techniques show that avoiding dead blocks is critical to maximize performance and reduce misses. Adopting more sophisticated replacement schemes is beneficial only under severe process variation.

5.5.2   EMBEDDED DRAM

Embedded DRAM (or eDRAM) is a technology from IBM specifically targeted to address the density problem of SRAM without compromising speed. It is similar to DRAM but implemented using high speed logic-based technology instead of the traditional DRAM process. When the overhead of peripheral circuits are taken into account, a 30 MB eDRAM cache, when optimized for speed, requires less than one-third the area of an equally sized SRAM. It also has superior leakage characteristics, consuming only one-fifth of the leakage of an SRAM cache. IBM Power 7 is the first processor to make use of this technology to build a massive 30 MB NUCA L3-cache to feed its eight out-of-order cores [220].

The problem with any DRAM based technology is its refresh overhead, especially when process variation is taken into account. The use of high performance transistors (transistors with low V_t but high leakage) in eDRAM exacerbates this problem and reduces its refresh interval by three orders of magnitude compared to DRAM. Due to large size of eDRAM caches, micro-managing refresh at a line granularity as discussed earlier is not feasible. To address this problem, Wilkerson et al. [223] propose using ECC to reduce refresh overhead. Ideally, when a line is written, all the cells in the line should retain their values just until the next refresh cycle. However, due to process variation, some cells will lose their value much earlier. Instead of refreshing the entire line based on the worst case retention time, and avoiding failures, Wilkerson et al. propose employing powerful error correcting codes to tolerate failed cells. Although, majority of failed lines can be corrected with simple SECDED, it cannot guarantee fault free operation for all the lines in a cache. Depending upon how aggressively we refresh cache lines, multi-bit errors can be non-trivial. Further, an error unrecognized or uncorrectable by SECDED will either crash the application or can lead to silent data corruption. Hence, a more powerful BCH code is employed, which can correct up to five errors and detect six errors. The primary drawback of BCH is its high decoding complexity to correct multi-bit errors. To avoid long latency penalty, the paper proposes splitting the detection process into two steps, quick-ECC, and Hi-ECC (or high latency ECC). All the lines with zero or one failed bits are corrected quickly by quick-ECC, and Hi-ECC is limited to those lines with multiple failures. Also, cache lines that are prone to multiple failures are disabled to further reduce latency penalty. With these optimizations, they achieve a 93% reduction in refresh power compared to a typical eDRAM implementation.

5.5.3   NON-VOLATILE MEMORIES

Most emerging technologies are non-volatile and are not charge based. Instead, they rely on varying resistance or magnetic properties of a cell to store values. Hence, there is no problem of leakage current as in DRAM or SRAM. Table 5.1 shows a few promising memory technologies, their latency, power, area, and endurance characteristics. It can be noticed that most of them have poor endurance property compared to SRAM or DRAM and their variants. Also, the overhead of write, both in terms of delay and power is significantly higher compare to reads. Techniques to extend lifetime, hide write delay, and reduce write power are active areas of research. Here, we briefly mention a couple of recent efforts to employ potential disruptive technologies for on-chip caches. Interested readers are referred to Qureshi et al.'s synthesis lecture on phase change memory for more recent work in the area [14].

Table 5.1: Memory Technology Comparison

Parameters	SRAM	DRAM	eDRAM	NAND FLASH	PCM	MRAM
Cell size	$120F^2$	$6 - 8F^2$	$16F^2$	$4 - 6F^2$	$4 - 16F^2$	$16 - 40F^2$
Read time	very fast	fast	faster than DRAM	slow	slower than DRAM	fast
Write time	very fast	fast	faster than DRAM	much slower than PCM	very slow	slow
Endurance	$> 10^{16}$	$> 10^{16}$	$> 10^{16}$	10^5	10^8	10^5
Non-Volatile	no	no	no	yes	yes	yes
Write power	very low	low	low	very high	high	high

3D Stacked MRAM, Sun et al., HPCA 2009

One of the key challenges in embracing a new technology is the cost of integrating it into the existing manufacturing process. Magnetoresistive-RAM (MRAM) stores values by changing the magnetic field in each cell. The technology requires special magnetic plates and fabrication steps different from the conventional CMOS process. To address this, Sun et al. [215] explore a 3D stacked design in which MRAM is stacked on top of a traditional CMOS processor. It also shows that MRAM's high write latency and write energy preclude it from being a clear winner for high level caches such as L2, especially for memory intensive applications. The paper proposes using a large write buffer and special scheduling policies that prioritize reads over writes to overcome the delay problem. To reduce the number of writes to MRAM (and hence power), they propose a hybrid cache for L2 consisting of both SRAM and MRAM banks. Cache lines that get frequently written are either placed in the SRAM bank or migrated to the SRAM bank. Their analysis shows promising results for a single threaded workload. However, as discussed earlier, this solution can also result in performance loss in CMPs where even writes are often on the critical path. Further, it may be possible to achieve the effect of the hybrid cache by having a small L2 SRAM cache and restricting MRAM for a lower level L3 cache. The next paper by Wu et al. [214] presents a detailed design space exploration on hybrid caches clarifying some of these options.

Hybrid Cache, Wu et al., ISCA 2009

Wu et al. [214] show how different technologies fare against each other in a hybrid cache architecture. They perform a detailed power and performance study of three different cache architectures, and for each, they consider caches built with SRAM, MRAM, PCM, and eDRAM.

Their first architecture (LHCA) models a traditional cache hierarchy with three levels: a private L1 made out of SRAM, a private L2, also made out of SRAM, and a shared L3, which can be any of the above storage technologies. The L3 cache is sized such that the effective area is the same for all the configurations they evaluate.

Their second architecture (RHCA) models a hybrid L2 cache, in which both L2 and L3 in the original design are combined into a single multi-banked cache. Each bank still retains its cell type making the cache hybrid. Since SRAM banks are always faster, it is beneficial to maximize the number of requests served by them. Hence, the paper proposes a migration policy to move the frequently used blocks from slower banks to SRAM banks. Their policy is most similar to D-NUCA but differs in when to swap blocks between banks. Their mapping scheme is similar to D-NUCA where each set is distributed across multiple banks (called bank set) to facilitate swapping. To search for a block, a request is sent to all the banks in the bank set. The criticality of cache lines in slow banks is determined by tracking the number of accesses to each cache line. This is done by maintaining a hardware saturating counter for every cache line. In fast SRAM banks, instead of a counter, a sticky bit is maintained for each cache line that gets set whenever a cache line is accessed. A block in a slow bank is ready to move as soon as the most significant bit in the saturating counter is set. However, swapping is allowed only if the corresponding block in the fast bank has its sticky bit set to zero. Otherwise, instead of swapping, sticky bit is reset to zero. This alleviates the problem of two hot blocks getting swapped frequently between a slow and a fast bank.

The third architecture leverages 3D-stacking to dramatically increase the size. In the CMP architecture discussed above, a new die specifically designed for caches is stacked on top of the processor. Since stacked caches are typically targeted for a lower level, fast access time is not a critical factor, but high density can help reduce miss rate. PCM, with its relatively low area requirement, is employed as the last level cache in all their 3D configurations. Similar to the first two architectures discussed, a 3D stacked design with various combinations of LHCA and RCHA are considered. This includes two RHCA combinations: (1) a hybrid L2 cache consisting of SRAM and eDRAM, and a stacked PCM for L3, and (2) a hybrid L2 cache consisting of SRAM, eDRAM, and PCM.

Here are some of the key take away points of their study. In an LHCA design, having a large L2 SRAM cache before an L3 MRAM cache is effective in filtering a significant number of writes to MRAM. In spite of MRAM's poor write power characteristics, an LHCA design with MRAM ended up with the best power value. With respect to performance, having an eDRAM L3 cache is more beneficial compared to MRAM, although the performance difference between them is very small. Between RHCA and the best performing LHCA, RHCA achieves better performance, but the difference is less than 2%. In 3D designs, a combination of an RHCA L2 (made of SRAM and eDRAM) and a PCM L3 achieves the best performance. Again, the performance of 3D LHCA is very close. In all likelihood, a complicated searching procedure and the bookkeeping required for RHCA is hard to justify compared to a much simpler LHCA design.

Summary

Cache architectures for emerging technologies continue to be an active area of research with large room for improvement. For instance, most upcoming technologies are non-volatile, but all existing proposals are oblivious to this. Many reliability techniques that previously required operating system or BIOS support can be performed completely in hardware within a processor since the data is persistent. Further, varying the characteristics of emerging NVMs to make them suitable for CMP caches as well as main memories is an interesting research direction. Although many previous studies have explored the benefit of stacking memory on processor [224], the effect of emerging technologies on such designs remains unresolved.

CHAPTER 6

Concluding Remarks

Even after decades of cache research, it is amazing that researchers continue to make significant advancements. The past decade has seen major innovations: the notion of non-uniform cache access, organizations that blur the boundary between shared and private, closer integration of cache and network, creative policies for insertion, replacement, and partitioning. There continues to be a need for better policies. There is a wide gap between state-of-the-art cache management and the performance of optimal cache management schemes (for example, Belady's OPT algorithm [86]). Nearly half the cache blocks in a large cache will never be accessed before eviction. The oracular bar is therefore set very high, and we'll continue to take small steps towards it. The key challenge is to achieve these small steps with techniques that are energy-efficient and not overly complex.

In the coming decade, we will not only continue to see improvements in on-chip caching, we will see many of these caching techniques being applied to new levels of the memory hierarchy. As process technologies continue to improve, we will see new artifacts within the memory hierarchy: new cells, new problems, 3D stacking, off-chip hierarchies with DRAM and NVM, non-uniform off-chip interconnect delays, and non-uniform queuing delays to go off-chip, to name a few. This highlights the importance of understanding emerging technology phenomena and the reason for including a chapter on technology issues in this book. We expect to see significant effort in the area of data management within the off-chip hierarchy. While the off-chip hierarchy has remained relatively simple in the past, future hierarchies will likely consist of combinations of different technologies: some NVM technology (PCM, STT-RAM) that provides high capacity and possibly some SRAM/DRAM-like technology that provides low-latency access. Given that DDRx electrical memory buses will only support one or two access points (DIMMs), some form of buffered off-chip network (for example, better incarnations of FB-DIMM) will be required to support the high memory capacity needs of large-scale multi-core multi-socket systems. The off-chip hierarchy will therefore start to resemble the networked L1-L2 hierarchies that are the focus of this book. They will also be constrained by the nuances of these new technologies, such as, the limited write endurance of PCM, the existence of row buffers or the significant costs of queuing delays. Access to these off-chip memories is expected to contribute greatly to overall system latency and energy. Many of the logical caching policies covered in this book will have to be adapted for off-chip data management efforts.

It may even be possible to devise co-ordinated mechanisms for on-chip caching and off-chip access that reduce overall system latency and energy. One recent paper tries to optimize memory row buffer hit rates by looking in the on-chip cache for blocks that are dead and that can be eagerly evicted [225]. Another recent paper tries to identify hot ranks in the memory system and modifies the on-chip cache replacement policy to de-prioritize eviction of blocks destined for hot ranks [226]. We therefore expect this area of co-ordinated caching and memory access mechanisms to receive much attention in the coming years.

There is clearly no shortage of interesting future work possibilities in the areas of on- and off-chip caching for large-scale multi-core systems.

Bibliography

[1] C. Kim, D. Burger, and S. Keckler. An Adaptive, Non-Uniform Cache Structure for Wire-Dominated On-Chip Caches. In *Proceedings of ASPLOS*, 2002. DOI: 10.1145/605397.605420 10, 11, 12, 15, 16, 17, 19, 21, 22

[2] B.M. Beckmann and D.A. Wood. Managing Wire Delay in Large Chip-Multiprocessor Caches. In *Proceedings of MICRO-37*, December 2004. DOI: 10.1109/MICRO.2004.21 8, 16, 19, 90, 96

[3] N. Hardavellas, M. Ferdman, B. Falsafi, and A. Ailamaki. Reactive NUCA: Near-Optimal Block Placement And Replication In Distributed Caches. In *Proceedings of ISCA*, 2009. DOI: 10.1145/1555754.1555779 7, 31, 32

[4] N. Muralimanohar, R. Balasubramonian, and N. Jouppi. Optimizing NUCA Organizations and Wiring Alternatives for Large Caches with CACTI 6.0. In *Proceedings of MICRO*, 2007. DOI: 10.1109/MICRO.2007.33 89, 90, 91

[5] Y. Jin, E. J. Kim, and K. H. Yum. A Domain-Specific On-Chip Network Design for Large Scale Cache Systems. In *Proceedings of HPCA*, 2007. DOI: 10.1109/HPCA.2007.346209 95, 100

[6] Z. Guz, I. Keidar, A. Kolodny, and U. Weiser. Utilizing Shared Data in Chip Multiprocessors with the Nahalal Architecture. In *Proceedings of SPAA*, June 2008. DOI: 10.1145/1378533.1378535 18, 96, 100

[7] N. Muralimanohar and R. Balasubramonian. Interconnect Design Considerations for Large NUCA Caches. In *Proceedings of ISCA*, 2007. DOI: 10.1145/1250662.1250708 8, 16, 53, 89, 97, 98, 100

[8] A. Udipi, N. Muralimanohar, and R. Balasubramonian. Towards Scalable, Energy-Efficient, Bus-Based On-Chip Networks. In *Proceedings of HPCA*, 2010. DOI: 10.1109/HPCA.2010.5416639 99, 100

[9] N. Madan, L. Zhao, N. Muralimanohar, A. Udipi, R. Balasubramonian, R. Iyer, S. Makineni, and D. Newell. Optimizing Communication and Capacity in a 3D Stacked Reconfigurable Cache Hierarchy. In *Proceedings of HPCA*, 2009. DOI: 10.1109/HPCA.2009.4798261 29, 108, 109

[10] B. Jacob. *The Memory System: You Can't Avoid It, You Can't Ignore It, You Can't Fake it*. Morgan & Claypool Synthesis Lectures on Computer Architecture, 2009. DOI: 10.2200/S00201ED1V01Y200907CAC007 xii

[11] N. Jerger and L. Peh. *On-Chip Networks*. Morgan & Claypool Synthesis Lectures on Computer Architecture, 2009. DOI: 10.2200/S00209ED1V01Y200907CAC008 xii, 81, 100

[12] A. Gonzalez, F. Latorre, and G. Magklis. *Processor Microarchitecture: An Implementation Perspective*. Morgan & Claypool Synthesis Lectures on Computer Architecture, 2010. DOI: 10.2200/S00309ED1V01Y201011CAC012 xii

[13] D. Sorin, M. Hill, and D. Wood. *A Primer on Memory Consistency and Cache Coherence*. Morgan & Claypool Synthesis Lectures on Computer Architecture, 2011. DOI: 10.2200/S00346ED1V01Y201104CAC016 xii

[14] M. Qureshi, S. Gurumurthi, and B. Rajendran. *Phase Change Memory: From Devices to Systems*. Morgan & Claypool Synthesis Lectures on Computer Architecture, 2011. xiii, 114

[15] S. Kaxiras and M. Martonosi. *Computer Architecture Techniques for Power-Efficiency*. Morgan & Claypool Synthesis Lectures on Computer Architecture, 2008. DOI: 10.2200/S00119ED1V01Y200805CAC004 xiii

[16] David E. Culler and Jaswinder Pal Singh. *Parallel Computer Architecture: A Hardware/Software Approach*. Morgan Kaufmann Publishers, 1999. 2

[17] J. L. Hennessy and D. A. Patterson. *Computer Architecture: A Quantitative Approach*. Elsevier, 4th edition, 2007. 2

[18] A. Jaleel, M. Mattina, and B. Jacob. Last Level Cache (LLC) Performance of Data Mining Workloads on a CMP – A Case Study of Parallel Bioinformatics Workloads. In *Proceedings of HPCA*, 2006. DOI: 10.1109/HPCA.2006.1598115 6

[19] C. Bienia, S. Kumar, and K. Li. PARSEC vs. SPLASH-2: A Quantitative Comparison of Two Multithreaded Benchmark Suites on Chip-Multiprocessors. In *Proceedings of IISWC*, 2008. DOI: 10.1109/IISWC.2008.4636090 7

[20] B. Beckmann, M. Marty, and D. Wood. ASR: Adaptive Selective Replication for CMP Caches. In *Proceedings of MICRO*, 2006. DOI: 10.1109/MICRO.2006.10 7, 24, 36, 38, 41, 51

[21] L. A. Barroso, K. Gharachorloo, R. McNamara, A. Nowatzyk, S. Qadeer, B. Sano, S. Smith, R. Stets, and B. Verghese. Piranha: A Scalable Architecture Based on Single-Chip Multiprocessing. In *Proceedings of ISCA-27*, pages 282–293, June 2000. DOI: 10.1145/342001.339696 8, 19, 81

[22] J. Huh, C. Kim, H. Shafi, L. Zhang, D. Burger, and S. Keckler. A NUCA Substrate for Flexible CMP Cache Sharing. In *Proceedings of ICS-19*, June 2005. DOI: 10.1109/TPDS.2007.1091 8, 18, 19, 22, 41, 90, 96

[23] Z. Chishti, M. Powell, and T.N. Vijaykumar. Optimizing Replication, Communication, and Capacity Allocation in CMPs. In *Proceedings of ISCA-32*, June 2005. DOI: 10.1145/1080695.1070001 13, 21, 22, 24

[24] Z. Chishti, M. Powell, and T.N. Vijaykumar. Distance Associativity for High-Performance Energy-Efficient Non-Uniform Cache Architectures. In *Proceedings of MICRO-36*, December 2003. DOI: 10.1109/MICRO.2003.1253183 13, 20, 21, 22, 64

[25] A. Jaleel, E. Borch, M. Bhandaru, S. Steely, and J. Emer. Achieving Non-Inclusive Cache Performance with Inclusive Caches – Temporal Locality Aware (TLA) Cache Management Policies. In *Proceedings of MICRO*, 2010. DOI: 10.1109/MICRO.2010.52 13, 63

[26] C. Liu, A. Sivasubramaniam, and M. Kandemir. Organizing the Last Line of Defense before Hitting the Memory Wall for CMPs. In *Proceedings of HPCA*, February 2004. DOI: 10.1109/HPCA.2004.10017 19, 20, 41, 44

[27] H. Dybdahl and P. Stenstrom. An Adaptive Shared/Private NUCA Cache Partitioning Scheme for Chip Multiprocessors. In *Proceedings of HPCA*, 2007. DOI: 10.1109/HPCA.2007.346180 19, 41, 44

[28] M. Qureshi, D. Thompson, and Y. Patt. The V-Way Cache: Demand Based Associativity via Global Replacement. In *Proceedings of ISCA*, 2005. DOI: 10.1145/1080695.1070015 21, 64

[29] D. Sanchez and C. Kozyrakis. The ZCache: Decoupling Ways and Associativity. In *Proceedings of MICRO*, 2010. DOI: 10.1109/MICRO.2010.20 21, 66

[30] R. Ricci, S. Barrus, D. Gebhardt, and R. Balasubramonian. Leveraging Bloom Filters for Smart Search Within NUCA Caches. In *Proceedings of the 7th Workshop on Complexity-Effective Design, held in conjunction with ISCA-33*, June 2006. 22

[31] Burton Bloom. Space/time trade-offs in hash coding with allowable errors, July 1970. DOI: 10.1145/362686.362692 22

[32] J. Merino, V. Puente, P. Prieto, and J. Gregorio. SP-NUCA: A Cost Effective Dynamic Non-Uniform Cache Architecture. *Computer Architecture News*, 2008. DOI: 10.1145/1399972.1399973 25

[33] J. Merino, V. Puente, and J. Gregorio. ESP-NUCA: A Low-Cost Adaptive Non-Uniform Cache Architecture. In *Proceedings of HPCA*, 2010. DOI: 10.1109/HPCA.2010.5416641 25

[34] R. E. Kessler and Mark D. Hill. Page Placement Algorithms for Large Real-Indexed Caches. *ACM Trans. Comput. Syst.*, 10(4), 1992. DOI: 10.1145/138873.138876 25

[35] R. Chandra, S. Devine, B. Verghese, A. Gupta, and M. Rosenblum. Scheduling and Page Migration for Multiprocessor Compute Servers. In *Proceedings of ASPLOS*, 1994. DOI: 10.1145/381792.195485 25

[36] J. Corbalan, X. Martorell, and J. Labarta. Page Migration with Dynamic Space-Sharing Scheduling Policies: The case of SGI 02000. *International Journal of Parallel Programming*, 32(4), 2004. DOI: 10.1023/B:IJPP.0000035815.13969.ec 25, 27

[37] B. Verghese, S. Devine, A. Gupta, and M. Rosenblum. Operating system support for improving data locality on CC-NUMA compute servers. *SIGPLAN Not.*, 31(9), 1996. DOI: 10.1145/248209.237205 25

[38] R.P. LaRowe, J.T. Wilkes, and C.S. Ellis. Exploiting Operating System Support for Dynamic Page Placement on a NUMA Shared Memory Multiprocessor. In *Proceedings of PPOPP*, 1991. DOI: 10.1145/109626.109639 25, 27

[39] R.P. LaRowe and C.S. Ellis. Experimental Comparison of Memory Management Policies for NUMA Multiprocessors. Technical report, 1990. DOI: 10.1145/118544.118546 25, 27

[40] R.P. LaRowe and C.S. Ellis. Page Placement policies for NUMA multiprocessors. *J. Parallel Distrib. Comput.*, 11(2), 1991. DOI: 10.1016/0743-7315(91)90117-R 25, 27

[41] S. Cho and L. Jin. Managing Distributed, Shared L2 Caches through OS-Level Page Allocation. In *Proceedings of MICRO*, 2006. DOI: 10.1109/MICRO.2006.31 25, 28, 29, 39, 41, 49, 51

[42] M. Awasthi, K. Sudan, R. Balasubramonian, and J. Carter. Dynamic Hardware-Assisted Software-Controlled Page Placement to Manage Capacity Allocation and Sharing within Large Caches. In *Proceedings of HPCA*, 2009. 28, 29, 33, 34, 39, 41, 49, 51, 52

[43] L. Jin and S. Cho. SOS: A Software-Oriented Distributed Shared Cache Management Approach for Chip Multiprocessors. In *Proceedings of PACT*, 2009. DOI: 10.1109/PACT.2009.14 28

[44] M. Marty and M. Hill. Virtual Hierarchies to Support Server Consolidation. In *Proceedings of ISCA*, 2007. DOI: 10.1145/1250662.1250670 29

[45] M. Chaudhuri. PageNUCA: Selected Policies For Page-Grain Locality Management In Large Shared Chip-Multiprocessor Caches. In *Proceedings of HPCA*, 2009. DOI: 10.1109/HPCA.2009.4798258 30, 33, 34, 39, 52

[46] J. Lin, Q. Lu, X. Ding, Z. Zhang, X. Zhang, and P. Sadayappan. Gaining Insights into Multicore Cache Partitioning: Bridging the Gap between Simulation and Real Systems. In *Proceedings of HPCA*, 2008. DOI: 10.1109/HPCA.2008.4658653 33, 41, 43, 51

[47] J. Lin, Q. Lu, X. Ding, Z. Zhang, X. Zhang, and P. Sadayappan. Enabling Software Management for Multicore Caches with a Lightweight Hardware Support. In *Proceedings of Supercomputing*, 2009. DOI: 10.1145/1654059.1654074 34, 41, 51

[48] J. Chang and G. Sohi. Co-Operative Caching for Chip Multiprocessors. In *Proceedings of ISCA*, 2006. DOI: 10.1145/1150019.1136509 35, 36, 37, 38, 49, 99

[49] E. Speight, H. Shafi, L. Zhang, and R. Rajamony. Adaptive Mechanisms and Policies for Managing Cache Hierarchies in Chip Multiprocessors. In *Proceedings of ISCA*, 2005. DOI: 10.1109/ISCA.2005.8 36, 38

[50] E. Herrero, J. Gonzalez, and R. Canal. Distributed Cooperative Caching. In *Proceedings of PACT*, 2008. DOI: 10.1145/1454115.1454136 37, 38

[51] E. Herrero, J. Gonzalez, and R. Canal. Elastic Cooperative Caching: An Autonomous Dynamically Adaptive Memory Hierarchy for Chip Multiprocessors. In *Proceedings of ISCA*, 2010. DOI: 10.1145/1816038.1816018 38

[52] M. K. Qureshi. Adaptive Spill-Receive for Robust High-Performance Caching in CMPs. In *Proceedings of HPCA*, 2009. 38, 41, 51

[53] Moinuddin K. Qureshi, Aamer Jaleel, Yale N. Patt, Simon C. Steely, and Joel Emer. Adaptive Insertion Policies for High Performance Caching. In *Proceedings of ISCA*, 2007. DOI: 10.1145/1250662.1250709 38, 43, 45, 58, 66

[54] H. Lee, S. Cho, and B. Childers. CloudCache: Expanding and Shrinking Private Caches. In *Proceedings of HPCA*, 2011. DOI: 10.1109/HPCA.2011.5749731 39

[55] H. Lee, S. Cho, and B. Childers. StimulusCache: Boosting Performance of Chip Multiprocessors with Excess Cache. In *Proceedings of HPCA*, 2010. DOI: 10.1109/HPCA.2010.5416644 39

[56] S. Srikantaiah, E. Kultursay, T. Zhang, M. Kandemir, M. Irwin, and Y. Xie. MorphCache: A Reconfigurable Adaptive Multi-Level Cache Hierarchy for CMPs. DOI: 10.1109/HPCA.2011.5749732 39

[57] T. Yeh and G. Reinman. Fast and Fair: Data-Stream Quality of Service. In *Proceedings of CASES*, 2005. DOI: 10.1145/1086297.1086328 42, 48, 52

[58] L. Hsu, S. Reinhardt, R. Iyer, and S. Makineni. Communist, Utilitarian, and Capitalist Cache Policies on CMPs: Caches as a Shared Resource. In *Proceedings of PACT*, 2006. DOI: 10.1145/1152154.1152161 42, 52

[59] M. Qureshi, D. Lynch, O. Mutlu, and Y. Patt. A Case for MLP-Aware Cache Replacement. In *Proceedings of ISCA*, 2006. DOI: 10.1109/ISCA.2006.5 43, 56

[60] M. Qureshi and Y. Patt. Utility-Based Cache Partitioning: A Low-Overhead, High-Performance, Runtime Mechanism to Partition Shared Caches. In *Proceedings of MICRO*, 2006. DOI: 10.1109/MICRO.2006.49 44, 48, 50, 54

[61] G.E. Suh, L. Rudolph, and S. Devadas. Dynamic Partitioning of Shared Cache Memory. *J. Supercomput.*, 28(1), 2004. DOI: 10.1023/B:SUPE.0000014800.27383.8f 44, 48

[62] X. Lin and R. Balasubramonian. Refining the Utility Metric for Utility-Based Cache Partitioning. In *Proceedings of WDDD*, 2011. 44

[63] G. Suo, X. Yang, G. Liu, J. Wu, K. Zeng, B. Zhang, and Y. Lin. IPC-Based Cache Partitioning: An IPC-Oriented Dynamic Shared Cache Partitioning Mechanism. In *Proceedings of ICHIT*, 2008. DOI: 10.1109/ICHIT.2008.164 44

[64] A. Jaleel, W. Hasenplaugh, M. Qureshi, J. Sebot, Jr. S. Steely, and J. Emer. Adaptive Insertion Policies For Managing Shared Caches. In *Proceedings of PACT*, 2008. DOI: 10.1145/1454115.1454145 45, 50

[65] Y. Xie and G. Loh. PIPP: Promotion/Insertion Pseudo-Partitioning of Multi-Core Shared Caches. In *Proceedings of ISCA*, 2009. DOI: 10.1145/1555815.1555778 46, 59, 61

[66] W. Liu and D. Yeung. Using Aggressor Thread Information to Improve Shared Cache Management for CMPs. In *Proceedings of PACT*, 2009. DOI: 10.1109/PACT.2009.13 47

[67] R. Balasubramonian, S. Dwarkadas, and D.H. Albonesi. Dynamically Managing the Communication-Parallelism Trade-Off in Future Clustered Processors. In *Proceedings of ISCA-30*, pages 275–286, June 2003. DOI: 10.1145/871656.859650 47

[68] S. Dropsho, A. Buyuktosunoglu, R. Balasubramonian, D. H. Albonesi, S. Dwarkadas, G. Semeraro, G. Magklis, and M. L. Scott. Integrating Adaptive On-Chip Storage Structures for Reduced Dynamic Power. In *Proceedings of the 11th International Conference on Parallel Architectures and Compilation Techniques (PACT)*, pages 141–152, September 2002. DOI: 10.1109/PACT.2002.1106013 48

[69] N. Rafique, W. Lim, and M. Thottethodi. Architectural Support for Operating System Driven CMP Cache Management. In *Proceedings of PACT*, 2006. DOI: 10.1145/1152154.1152160 48

[70] J. Chang and G. Sohi. Co-Operative Cache Partitioning for Chip Multiprocessors. In *Proceedings of ICS*, 2007. DOI: 10.1145/1274971.1275005 49

[71] S. Srikantaiah, M. Kandemir, and M. Irwin. Adaptive Set Pinning: Managing Shared Caches in Chip Multiprocessors. In *Proceedings of ASPLOS*, 2008. DOI: 10.1145/1353534.1346299 49

[72] T. Sherwood, B. Calder, and J. Emer. Reducing Cache Misses Using Hardware and Software Page Placement. In *Proceedings of SC*, 1999. DOI: 10.1145/305138.305189 49, 64

[73] X. Jiang, A. Mishra, L. Zhao, R. Iyer, Z. Fang, S. Srinivasan, S. Makineni, P. Brett, and C. Das. ACCESS: Smart Scheduling for Asymmetric Cache CMPs. In *Proceedings of HPCA*, 2011. DOI: 10.1109/HPCA.2011.5749757 50

[74] D. Chandra, F. Guo, S. Kim, and Y. Solihin. Predicting inter-thread cache contention on a chip-multiprocessor architecture. In *Proceedings of HPCA-11*, February 2005. DOI: 10.1109/HPCA.2005.27 50

[75] D. Tam, R. Azimi, L. Soares, and M. Stumm. RapidMRC: Approximating L2 Miss Rate Curves on Commodity Systems for Online Optimizations. In *Proceedings of ASPLOS*, 2009. DOI: 10.1145/1508284.1508259 51, 52

[76] S. Zhuravlev, S. Blagodurov, and A. Fedorova. Addressing Shared Resource Contention in Multicore Processors via Scheduling. In *Proceedings of ASPLOS*, 2010. DOI: 10.1145/1735971.1736036 51

[77] R. Iyer, L. Zhao, F. Guo, R. Illikkal, D. Newell, Y. Solihin, L. Hsu, and S. Reinhardt. QoS Policies and Architecture for Cache/Memory in CMP Platforms. In *Proceedings of SIGMET-RICS*, 2007. DOI: 10.1145/1254882.1254886 52, 54

[78] R. Iyer. CQoS: A Framework for Enabling QoS in Shared Caches of CMP Platforms. In *Proceedings of ICS*, 2004. DOI: 10.1145/1006209.1006246 53

[79] K. Varadarajan, S. Nandy, V. Sharda, A. Bharadwaj, R. Iyer, S. Makineni, and D. Newell. Molecular Caches: A Caching Structure for Dynamic Creation of Application-Specific Heterogeneous Cache Regions. In *Proceedings of MICRO*, 2006. DOI: 10.1109/MICRO.2006.38 53

[80] F. Guo, Y. Solihin, L. Zhao, and R. Iyer. A Framework for Providing Quality of Service in Chip Multi-Processors. In *Proceedings of MICRO*, 2007. DOI: 10.1109/MICRO.2007.6 53, 54

[81] Li Zhao, Ravi Iyer, Ramesh Illikkal, Jaideep Moses, Srihari Makineni, and Don Newell. CacheScouts: Fine-Grain Monitoring of Shared Caches in CMP Platforms. In *Proceedings of PACT*, 2007. DOI: 10.1109/PACT.2007.19 54

[82] S. Srikantaiah, M. Kandemir, and Q. Wang. SHARP Control: Controlled Shared Cache Management in Chip Multiprocessors. In *Proceedings of MICRO*, 2009. DOI: 10.1145/1669112.1669177 54

[83] S. Srikantaiah, R. Das, A. Mishra, C. Das, and M. Kandemir. A Case for Integrated Processor-Cache Partitioning in Chip Multiprocessors. In *Proceedings of Supercomputing*, 2009. DOI: 10.1145/1654059.1654066 54

[84] K. Nesbit, J. Laudon, and J. E. Smith. Virtual private caches. In *Proceedings ISCA*, 2007. DOI: 10.1145/1250662.1250671 55

[85] S. Kim, D. Chandra, and Y. Solihin. Fair Cache Sharing and Partitioning in a Chip Multiprocessor Architecture. In *Proceedings PACT*, 2004. DOI: 10.1109/PACT.2004.1342546 55

[86] L. Belady. A Study of Replacement Algorithms for a Virtual-Storage Computer. *IBM Systems Journal*, 1966. DOI: 10.1147/sj.52.0078 56, 58, 117

[87] J. Jeong and M. Dubois. Optimal Replacements in Caches with Two Miss Costs. In *Proceedings of SPAA*, 1999. DOI: 10.1145/305619.305636 56

[88] J. Jeong and M. Dubois. Cost-Sensitive Cache Replacement Algorithms. In *Proceedings of HPCA*, 2003. DOI: 10.1109/HPCA.2003.1183550 56, 57

[89] W. Wong and J. Baer. Modified LRU Policies for Improving Second-Level Cache Behavior. In *Proceedings of HPCA*, 2000. DOI: 10.1109/HPCA.2000.824338 56

[90] A. Jaleel, K. Theobald, S. Steely, and J. Emer. High Performance Cache Replacement Using Re-Reference Interval Prediction (RRIP). In *Proceedings of ISCA*, 2010. 56, 60

[91] S. Srinivasan, R. Ju, A. Lebeck, and C. Wilkerson. Locality vs. Criticality. In *Proceedings of ISCA-28*, pages 132–143, July 2001. DOI: 10.1109/ISCA.2001.937442 57

[92] R. Balasubramonian, V. Srinivasan, and S. Dwarkadas. Hot-and-Cold: Using Criticality in the Design of Energy-Efficient Caches. In *Workshop on Power-Aware Computer Systems, in conjunction with MICRO-36*, December 2003. 57

[93] R. Subramanian, Y. Smaragdakis, and G. Loh. Adaptive Caches: Effective Shaping of Cache Behavior to Workloads. In *Proceedings of MICRO-39*, December 2006. DOI: 10.1109/MICRO.2006.7 57

[94] G. Keramidas, P. Petoumenos, and S. Kaxiras. Cache Replacement based on Reuse-Distance Prediction. In *Proceedings of ICCD*, 2007. DOI: 10.1109/ICCD.2007.4601909 58

[95] K. Rajan and R. Govindarajan. Emulating Optimal Replacement with a Shepherd Cache. In *Proceedings of MICRO*, 2007. DOI: 10.1109/MICRO.2007.14 58, 59, 61, 72

[96] J. Zebchuk, S. Makineni, and D. Newell. Re-examining Cache Replacement Policies. In *Proceedings of ICCD*, 2008. DOI: 10.1109/ICCD.2008.4751933 59

[97] K. So and R. Rechtshaffen. Cache Operations by MRU Change. *IEEE Transactions on Computers*, 37(6), June 1988. DOI: 10.1109/12.2208 59

[98] D. Lee, J. Choi, J. Kim, S. Noh, S. Min, Y. Cho, and C. Kim. LRFU: A Spectrum of Policies that Subsumes the Least Recently Used and Least Frequently Used Policies. *IEEE Trans. on Computers*, 50(12), 2001. DOI: 10.1109/TC.2001.970573 60

[99] M. Chaudhuri. Pseudo-LIFO: The Foundation of a New Family of Replacement Policies for Last-level Caches. In *Proceedings of MICRO*, 2009. DOI: 10.1145/1669112.1669164 61, 62

[100] M. Kharbutli and Y. Solihin. Counter-Based Cache Replacement and Bypassing Algorithms. *IEEE Trans. on Computers*, 2008. DOI: 10.1109/TC.2007.70816 62, 67, 71, 72

[101] H. Liu, M. Ferdman, J. Huh, and D. Burger. Cache Bursts: A New Approach for Eliminating Dead Blocks and Increasing Cache Efficiency. In *Proceedings of MICRO*, 2008. DOI: 10.1109/MICRO.2008.4771793 62, 67, 71

[102] S. Khan, D. Jimenez, D. Burger, and B. Falsafi. Using Dead Blocks as a Virtual Victim Cache. In *Proceedings of PACT*, 2010. DOI: 10.1145/1854273.1854333 62, 65, 67, 72

[103] A. Basu, N. Kirman, M. Kirman, M. Chaudhuri, and J. Martinez. Scavenger: A New Last Level Cache Architecture with Global Block Priority. In *Proceedings of MICRO*, 2007. DOI: 10.1109/MICRO.2007.36 62

[104] N. Jouppi. Improving Direct-Mapped Cache Performance by the Addition of a Small Fully-Associative Cache and Prefetch Buffers. In *Proceedings of ISCA-17*, pages 364–373, May 1990. DOI: 10.1109/ISCA.1990.134547 62, 67, 68

[105] R. Manikantan, K. Rajan, and R. Govindarajan. NUcache: An Efficient Multicore Cache Organization Based on Next-Use Distance. In *Proceedings of HPCA*, 2011. DOI: 10.1145/1854273.1854356 62

[106] L. Soares, D. Tam, and M. Stumm. Reducing the Harmful Effects of Last-Level Cache Polluters with an OS-Level, Software-Only Pollute Buffer. In *Proceedings of MICRO*, 2008. DOI: 10.1109/MICRO.2008.4771796 63

[107] Intel. Intel Core i7 Processor. http://www.intel.com/products/processor/corei7/specifications.htm. 64

[108] AMD. AMD Athlon Processor and AMD Duron Processor with Full-Speed On-die L2 Cache, 2000. 64

[109] A. Seznec. A Case for Two-Way Skewed-Associative Caches. In *Proceedings of ISCA*, 1993. DOI: 10.1145/173682.165152 64, 66

[110] M. Kharbutli, K. Irwin, Y. Solihin, and J. Lee. Using Prime Numbers for Cache Indexing to Eliminate Conflict Misses. In *Proceedings of HPCA*, 2004. DOI: 10.1109/HPCA.2004.10015 64

[111] D. Rolan, B. Fraguela, and R. Doallo. Adaptive Line Placement with the Set Balancing Cache. In *Proceedings of MICRO*, 2009. DOI: 10.1145/1669112.1669178 65, 72

[112] C. Zhang. Balanced Cache: Reducing Conflict Misses of Direct-Mapped Caches. In *Proceedings of ISCA*, 2006. DOI: 10.1145/1150019.1136499 65

[113] D. Zhan, H. Jiang, and S. Seth. STEM: Spatiotemporal Management of Capacity for Intra-Core Last Level Caches. In *Proceedings of MICRO*, 2010. DOI: 10.1109/MICRO.2010.31 66

[114] T. Chen and J. Baer. Effective Hardware Based Data Prefetching for High Performance Processors. *IEEE Transactions on Computers*, 44(5):609–623, May 1995. DOI: 10.1109/12.381947 67

[115] A.J. Smith. Cache Memories. *Computing Surveys*, 14(4), 1982. DOI: 10.1145/356887.356892 67

[116] Alaa Alameldeen and David Wood. Interactions Between Compression and Prefetching in Chip Multiprocessors. In *Proceedings of HPCA*, 2007. DOI: 10.1109/HPCA.2007.346200 67, 75

[117] E. Ebrahimi, O. Mutlu, and Y. Patt. Techniques for Bandwidth-Efficient Prefetching of Linked Data Structures in Hybrid Prefetching Systems. In *Proceedings of HPCA*, 2009. 67, 70

[118] A. Lai, C. Fide, and B. Falsafi. Dead-Block Prediction and Dead-Block Correlating Prefetchers. In *Proceedings of ISCA*, 2001. DOI: 10.1145/379240.379259 67, 72

[119] D. Wood, M. Hill, and R. Kessler. A Model for Estimating Trace-Sample Miss Ratios. In *Proceedings of SIGMETRICS*, 1991. DOI: 10.1145/107972.107981 67

[120] Z. Hu, S. Kaxiras, and M. Martonosi. Timekeeping in the Memory System: Predicting and Optimizing Memory Behavior. In *Proceedings of ISCA*, 2002. DOI: 10.1145/545214.545239 67

[121] M. Ferdman and B. Falsafi. Last-Touch Correlated Data Streaming. In *Proceedings of ISPASS*, 2007. DOI: 10.1109/ISPASS.2007.363741 67

[122] K. Flautner, N.S. Kim, S. Martin, D. Blaauw, and T. Mudge. Drowsy Caches: Simple Techniques for Reducing Leakage Power. In *Proceedings of ISCA*, 2002.
DOI: 10.1109/ISCA.2002.1003572 67

[123] J. Abella, A. Gonzalez, X. Vera, and M. O'Boyle. IATAC: A Smart Predictor to Turn-Off L2 Cache Lines. *ACM Trans. on Architecture and Code Optimization*, 2005.
DOI: 10.1145/1061267.1061271 67, 71

[124] S. Kaxiras, Z. Hu, and M. Martonosi. Cache Decay: Exploiting Generational Behavior to Reduce Cache Leakage Power. In *Proceedings of ISCA*, 2001. DOI: 10.1109/ISCA.2001.937453 67

[125] Z. Hu, M. Martonosi, and S. Kaxiras. TCP: Tag Correlating Prefetchers. In *Proceedings of HPCA*, 2003. DOI: 10.1109/HPCA.2003.1183549 67

[126] K. Nesbit and J. E. Smith. Data Cache Prefetching Using a Global History Buffer. In *Proceedings HPCA*, 2004. DOI: 10.1109/HPCA.2004.10030 67

[127] R. Manikantan and R. Govindarajan. Focused Prefetching: Performance Oriented Prefetching Based on Commit Stalls. In *Proceedings of ICS*, 2008. DOI: 10.1145/1375527.1375576 67

[128] H. Zhu, Y. Chen, and X. Sun. Timing Local Streams: Improving Timeliness in Data Prefetching. In *Proceedings of ICS*, 2010. DOI: 10.1145/1810085.1810110 67

[129] M. Grannaes, M. Jahre, and L. Natvig. Multi-Level Hardware Prefetching using Low Complexity Delta Correlating Prediction Tables with Partial Matching. In *Proceedings of HiPEAC*, 2010. DOI: 10.1007/978-3-642-11515-8_19 67

[130] V. Srinivasan, E. Davidson, and G. Tyson. A Prefetch Taxonomy. *IEEE Trans. on Computers*, February 2004. DOI: 10.1109/TC.2004.1261824 67

[131] L. Spracklen, Y. Chou, and S. Abraham. Effective Instruction Prefetching in Chip Multiprocessors for Modern Commercial Applications. In *Proceedings of HPCA*, 2005.
DOI: 10.1109/HPCA.2005.13 67

[132] T. Wenisch, S. Somogyi, N. Hardavellas, J. Kim, A. Ailamaki, and B. Falsafi. Temporal Streaming of Shared Memory. In *Proceedings of ISCA*, 2005. DOI: 10.1145/1080695.1069989 68

[133] T. Wenisch, M. Ferdman, A. Ailamaki, B. Falsafi, and A. Moshovos. Practical Off-Chip Meta-Data for Temporal Memory Streaming. In *Proceedings of HPCA*, 2009.
DOI: 10.1109/HPCA.2009.4798239 69

[134] M. Ferdman, T. Wenisch, A. Ailamaki, B. Falsafi, and A. Moshovos. Temporal Instruction Fetch Streaming. In *Proceedings of MICRO*, 2008. DOI: 10.1109/MICRO.2008.4771774 69

[135] S. Somogyi, T. Wenisch, A. Ailamaki, B. Falsafi, and A. Moshovos. Spatial Memory Streaming. In *Proceedings of ISCA*, 2006. DOI: 10.1145/1150019.1136508 69

[136] S. Somogyi, T. Wenisch, A. Ailamaki, and B. Falsafi. Spatio-Temporal Memory Streaming. In *Proceedings of ISCA*, 2009. DOI: 10.1145/1555754.1555766 69

[137] S. Srinath, O. Mutlu, H. Kim, and Y. Patt. Feedback Directed Prefetching: Improving the Performance and Bandwidth-Efficiency of Hardware Prefetchers. In *Proceedings of HPCA*, 2007. DOI: 10.1109/HPCA.2007.346185 70

[138] J.M. Tendler, S. Dodson, S. Fields, H. Le, and B. Sinharoy. Power4 System Microarchitecture. Technical report, Technical White Paper, IBM, October 2001. DOI: 10.1147/rd.461.0005 70

[139] I. Hur and C. Lin. Memory Prefetching Using Adaptive Stream Detection. In *Proceedings of MICRO*, 2006. DOI: 10.1109/MICRO.2006.32 70

[140] I. Hur and C. Lin. Feedback Mechanisms for Improving Probabilistic Memory Prefetching. In *Proceedings of HPCA*, 2009. DOI: 10.1109/HPCA.2009.4798282 70

[141] E. Ebrahimi, O. Mutlu, C. Lee, and Y. Patt. Coordinated Control of Multiple Prefetchers in Multi-Core Systems. In *Proceedings of MICRO*, 2009. DOI: 10.1145/1669112.1669154 70

[142] S. Khan, Y. Tian, and D. Jimenez. Dead Block Replacement and Bypass with a Sampling Predictor. In *Proceedings of MICRO*, 2010. 72

[143] C. Chen, S. Yang, B. Falsafi, and A. Moshovos. Accurate and Complexity-Effective Spatial Pattern Prediction. In *Proceedings of HPCA*, 2004. DOI: 10.1109/HPCA.2004.10010 73

[144] T. Johnson. *Run-time Adaptive Cache Management*. PhD thesis, University of Illinois, May 1998. 73

[145] S. Kumar and C. Wilkerson. Exploiting Spatial Locality in Data Caches Using Spatial Footprints. In *Proceedings of ISCA*, 1998. DOI: 10.1109/ISCA.1998.694794 73

[146] A. Seznec. Decoupled Sectored Caches: Conciliating Low Tag Implementation Cost. In *Proceedings of ISCA*, 1994. DOI: 10.1109/ISCA.1994.288133 73, 74

[147] P. Pujara and A. Aggarwal. Increasing the Cache Efficiency by Eliminating Noise. In *Proceedings of HPCA*, 2006. DOI: 10.1109/HPCA.2006.1598121 73

[148] P. Pujara and A. Aggarwal. Increasing Cache Capacity through Word Filtering. In *Proceedings of ICS*, 2007. DOI: 10.1145/1274971.1275002 73

[149] M. Ekman and P. Stenstrom. A Robust Main-Memory Compression Scheme. In *Proceedings of ISCA*, 2005. DOI: 10.1145/1080695.1069978 73

[150] B. Abali, H. Franke, S. Xiaowei, D. Poff, and T. Smith. Performance of Hardware Compressed Main Memory. In *Proceedings of HPCA*, 2001. DOI: 10.1109/HPCA.2001.903253 73

[151] R. Tremaine, P. Franaszek, J. Robinson, C. Schulz, T. Smith, M. Wazlowski, and P. Bland. IBM Memory Expansion Technology (MXT). *IBM Journal of Research and Development*, 45(2), 2001. DOI: 10.1147/rd.452.0271 73

[152] J. Yang, Y. Zhang, and R. Gupta. Frequent Value Compression in Data Caches. In *Proceedings of MICRO-33*, pages 258–265, December 2000. DOI: 10.1145/360128.360154 73

[153] Y. Zhang, J. Yang, and R. Gupta. Frequent Value Locality and Value-Centric Data Cache Design. In *Proceedings of ASPLOS*, 2000. DOI: 10.1090/S0002-9939-99-05318-6 73

[154] M. Qureshi, M. Suleman, and Y. Patt. Line Distillation: Increasing Cache Capacity by Filtering Unused Words in Cache Lines. In *Proceedings of HPCA*, 2007. DOI: 10.1109/HPCA.2007.346202 73, 75

[155] Alaa Alameldeen and David Wood. Adaptive Cache Compression for High-Performance Processors. In *Proceedings of ISCA*, 2004. DOI: 10.1145/1028176.1006719 74, 75

[156] J. Lee, W. Hong, and S. Kim. Design and Evaluation of a Selective Compressed Memory System. In *Proceedings of ICCD*, 1999. DOI: 10.1109/ICCD.1999.808424 74

[157] E. Hallnor and S. Reinhardt. A Unified Compressed Memory Hierarchy. In *Proceedings of HPCA*, 2005. DOI: 10.1109/HPCA.2005.4 74

[158] E. Hallnor and S. Reinhardt. A Fully Associative Software-Managed Cache Design. In *Proceedings of ISCA*, 2000. DOI: 10.1145/339647.339660 74

[159] X. Jiang, N. Madan, L. Zhao, M. Upton, R. Iyer, S. Makineni, D. Newell, Y. Solihin, and R. Balasubramonian. CHOP: Adaptive Filter-Based DRAM Caching for CMP Server Platforms. In *Proceedings of HPCA*, 2010. 75

[160] R. Ho, K.W. Mai, and M.A. Horowitz. The Future of Wires. *Proceedings of the IEEE*, Vol.89, No.4, April 2001. DOI: 10.1109/5.920580 79

[161] P. Kongetira, K. Aingaran, and K. Olukotun. Niagara: A 32-Way Multithreaded Sparc Processor. *IEEE Micro*, 25(2), 2005. DOI: 10.1109/MM.2005.35 81

[162] R. Kumar, V. Zyuban, and D. Tullsen. Interconnections in Multi-Core Architectures: Understanding Mechanisms, Overheads, and Scaling. In *Proceedings of ISCA*, 2005. DOI: 10.1109/ISCA.2005.34 81

[163] W.J. Dally and B. Towles. *Principles and Practices of Interconnection Networks*. Morgan Kaufmann, 1st edition, 2003. 81, 86

[164] W. Dally and B. Towles. Route Packets, Not Wires: On-Chip Interconnection Networks. In *Proceedings of DAC*, 2001. DOI: 10.1109/DAC.2001.156225 81

[165] P. Kundu. On-Die Interconnects for Next Generation CMPs. In *Workshop on On- and Off-Chip Interconnection Networks for Multicore Systems (OCIN)*, 2006. 83, 88

[166] Hang-Sheng Wang, Xinping Zhu, Li-Shiuan Peh, and Sharad Malik. Orion: A Power-Performance Simulator for Interconnection Networks. In *Proceedings of MICRO-35*, November 2002. DOI: 10.1109/MICRO.2002.1176258 83, 90

[167] H-S. Wang, L-S. Peh, and S. Malik. Power-Driven Design of Router Microarchitectures in On-Chip Networks. In *Proceedings of MICRO*, 2003. DOI: 10.1109/MICRO.2003.1253187 83, 89

[168] H.-S. Wang, L.-S. Peh, and S. Malik. A Power Model for Routers: Modeling Alpha 21364 and InfiniBand Routers. In *IEEE Micro, Vol 24, No 1*, January 2003. DOI: 10.1109/MM.2003.1179895 83, 88

[169] T. Moscibroda and O. Mutlu. A Case for Bufferless Routing in On-Chip Networks. In *Proceedings of ISCA*, 2009. DOI: 10.1145/1555754.1555781 83, 89

[170] R. Mullins, A. West, and S. Moore. Low-Latency Virtual-Channel Routers for On-Chip Networks. In *Proceedings of ISCA*, 2004. DOI: 10.1109/ISCA.2004.1310774 86

[171] J. Howard et al. A 48-Core IA-32 Message-Passing Processor with DVFS in 45nm CMOS. In *Proceedings of ISSCC*, 2010. DOI: 10.1109/JSSC.2010.2079450 88

[172] C. Nicopoulos, D. Park, J. Kim, V. Narayanan, M. Yousif, and C. Das. ViChaR: A Dynamic Virtual Channel Regulator for Network-on-Chip Routers. In *Proceedings of MICRO*, 2006. DOI: 10.1109/MICRO.2006.50 89

[173] J. Kim, C. Nicopoulos, D. Park, V. Narayanan, M. Yousif, and C. Das. A Gracefully Degrading and Energy-Efficient Modular Router Architecture for On-Chip Networks. In *Proceedings of ISCA*, 2006. DOI: 10.1109/ISCA.2006.6 89

[174] R. Ho. *On-Chip Wires: Scaling and Efficiency*. PhD thesis, Stanford University, August 2003. 89, 93

[175] W. J. Dally. Express Cubes: Improving the Performance of k-ary n-cube Interconnection Networks. *IEEE Transactions on Computers*, 40(9), 1991. DOI: 10.1109/12.83652 89

[176] A. Kumar, L. Peh, P. Kundu, and N. Jha. Express Virtual Channels: Towards the Ideal Interconnection Fabric. In *Proceedings of ISCA*, 2007. DOI: 10.1145/1273440.1250681 89

[177] J. Balfour and W. J. Dally. Design Tradeoffs for Tiled CMP On-Chip Networks. In *Proceedings of ICS*, 2006. DOI: 10.1145/1183401.1183430 89

[178] N. Muralimanohar, R. Balasubramonian, and N. Jouppi. Architecting Efficient Interconnects for Large Caches with CACTI 6.0. *IEEE Micro (Special Issue on Top Picks from Architecture Conferences)*, Jan/Feb 2008. DOI: 10.1109/MM.2008.2 89

[179] R. Ho, K. Mai, and M. Horowitz. Efficient On-Chip Global Interconnects. In *Proceedings of VLSI*, 2003. DOI: 10.1023/A:1023331803664 90, 93

[180] A. Udipi, N. Muralimanohar, and R. Balasubramonian. Non-Uniform Power Access in Large Caches with Low-Swing Wires. In *Proceedings of HiPC*, 2009. DOI: 10.1109/HIPC.2009.5433222 93

[181] M. Minzuno, K. Anjo, Y. Sume, M. Fukaishi, H. Wakabayashi, T. Mogami, T. Horiuchi, and M. Yamashina. Clock Distribution Networks with On-Chip Transmission Lines. In *Proceedings of the IEEE International Interconnect Technology Conference*, pages 3–5, 2000. 94

[182] J.D. Warnock, J.M. Keaty, J. Petrovick, J.G. Clabes, C.J. Kircher, B.L. Krauter, P.J. Restle, B.A. Zoric, and C.J. Anderson. The Circuit and Physical Design of the POWER4 Microprocessor. *IBM Journal of Research and Development*, 46(1):27–51, January 2002. DOI: 10.1147/rd.461.0027 94

[183] T. Xanthopoulos, D.W. Bailey, A.K. Gangwar, M.K. Gowan, A.K. Jain, and B.K. Prewitt. The Design and Analysis of the Clock Distribution Network for a 1.2GHz Alpha Microprocessor. In *Proceedings of the IEEE International Solid-State Circuits Conference*, pages 402–403, 2001. DOI: 10.1109/ISSCC.2001.912693 94

[184] R.T. Chang, N. Talwalkar, C.P. Yue, and S.S. Wong. Near Speed-of-Light Signaling Over On-Chip Electrical Interconnects. *IEEE Journal of Solid-State Circuits*, 38(5):834–838, May 2003. DOI: 10.1109/JSSC.2003.810060 94

[185] A. Deutsch. Electrical Characteristics of Interconnections for High-Performance Systems. *Proceedings of the IEEE*, 86(2):315–355, February 1998. DOI: 10.1109/5.659489 94

[186] B.M. Beckmann and D.A. Wood. TLC: Transmission Line Caches. In *Proceedings of MICRO-36*, December 2003. DOI: 10.1109/MICRO.2003.1253182 94, 100

[187] R. Manevich, I. Walter, I. Cidon, and A. Kolodny. Best of Both Worlds: A Bus-Enhanced NoC (BENoC). In *Proceedings of NOCS*, 2009. DOI: 10.1109/NOCS.2009.5071465 97, 100

[188] L. Cheng, N. Muralimanohar, K. Ramani, R. Balasubramonian, and J. Carter. Interconnect-Aware Coherence Protocols for Chip Multiprocessors. In *Proceedings of 33rd International Symposium on Computer Architecture (ISCA-33)*, pages 339–350, June 2006. DOI: 10.1109/ISCA.2006.23 97, 100

[189] E. Bolotin, Z. Guz, I. Cidon, R. Ginosar, and A. Kolodny. The Power of Priority: NoC based Distributed Cache Coherency. In *Proceedings of NOCS*, 2007. DOI: 10.1109/NOCS.2007.42 98, 100

[190] N. Eisley, L-S. Peh, and L. Shang. In-Network Cache Coherence. In *Proceedings of MICRO*, 2006. DOI: 10.1109/MICRO.2006.27 98, 100

[191] N. Eisley, L-S. Peh, and L. Shang. Leveraging On-Chip Networks for Cache Migration in Chip Multiprocessors. In *Proceedings of PACT*, 2008. DOI: 10.1145/1454115.1454144 99, 100

[192] N. Agarwal, L-S. Peh, and N. Jha. In-Network Snoop Ordering: Snoopy Coherence on Un-ordered Interconnects. In *Proceedings of HPCA*, 2009. DOI: 10.1109/HPCA.2009.4798238 99, 100

[193] N. Agarwal, L-S. Peh, and N. Jha. In-Network Coherence Filtering: Snoopy Coherence without Broadcasts. In *Proceedings of MICRO*, 2009. DOI: 10.1145/1669112.1669143 99, 100

[194] R. Das, S. Eachempati, A. K. Mishra, N. Vijaykrishnan, and C. R. Das. Design and Evaluation of Hierarchical On-Chip Network Topologies for Next Generation CMPs. In *Proceedings of HPCA*, 2009. 100

[195] S. Borkar. Designing Reliable Systems from Unreliable Components: The Challenges of Transistor Variability and Degradation. *IEEE Micro*, Nov/Dec 2005. DOI: 10.1109/MM.2005.110 102

[196] Y. B. Kim et al. Low Power 8T SRAM Using 32nm Independent Gate FinFET Technology. In *SOC Conference*, 2008. DOI: 10.1109/SOCC.2008.4641521 102

[197] S. M. Jahinuzzaman et al. A Soft Error Tolerant 10T SRAM Bit-Cell With DIfferential Read Capability. In *IEEE Transactions on Nuclear Science*, 2009. DOI: 10.1109/TNS.2009.2032090 102

[198] S. Borkar, T. Karnik, S. Narendra, J. Tschanz, A. Keshavarzi, and V. De. Parameter Variations and Impact on Circuits and Microarchitecture. In *Proceedings of DAC*, 2003. DOI: 10.1109/DAC.2003.1219020 102

[199] E. Humenay, D. Tarjan, and K. Skadron. 103

[200] ITRS. International Technology Roadmap for Semiconductors, 2009 Edition. 103

[201] S. Ozdemir, D. Sinha, G. Memik, J. Adams, and H. Zhou. Yield-Aware Cache Architectures. In *Proceedings of MICRO*, 2006. DOI: 10.1109/MICRO.2006.52 104

[202] D.H. Albonesi. Selective Cache Ways: On-Demand Cache Resource Allocation. 1999. DOI: 10.1109/MICRO.1999.809463 104, 105

[203] A. Das, B. Ozisikylmaz, S. Ozdemir, G. Memik, J. Zambreno, and A. Choudhary. Evaluating the Effects of Cache Redundancy on Profit. In *Proceedings of MICRO*, 2008. DOI: 10.1109/MICRO.2008.4771807 105

[204] R. C. Bose and D. K. Ray-Chaudhuri. On a Class of Error Correcting Binary Group Codes. In *Information and Control*, 1960. DOI: 10.1016/S0019-9958(60)90870-6 106

[205] J. Kim, N. Hardavellas, K. Mai, B. Falsafi, and J. C. Hoe. 106, 107

[206] D. H. Yoon and M. Erez. Memory Mapped ECC: Low-Cost Error Protection for Last Level Caches. In *Proceedings of ISCA*, 2009. DOI: 10.1145/1555815.1555771 107

[207] Y. Xie, G. Loh, B. Black, and K. Bernstein. Design Space Exploration for 3D Architectures. *ACM Journal of Emerging Technologies in Computing Systems*, 2(2):65–103, April 2006. DOI: 10.1145/1148015.1148016 108

[208] T. Kgil, S. D'Souza, A. Saidi, N. Binkert, R. Dreslinski, S. Reinhardt, K. Flautner, and T. Mudge. PicoServer: Using 3D Stacking Technology to Enable a Compact Energy Efficient Chip Multiprocessor. In *Proceedings of ASPLOS*, 2006. DOI: 10.1145/1168919.1168873 108

[209] F. Li, C. Nicopoulos, T. Richardson, Y. Xie, N. Vijaykrishnan, and M. Kandemir. Design and Management of 3D Chip Multiprocessors Using Network-in-Memory. In *Proceedings of ISCA-33*, June 2006. DOI: 10.1109/ISCA.2006.18 108

[210] C. C. Liu, I. Ganusov, M. Burtscher, and S. Tiwari. Bridging the Processor-Memory Performance Gap with 3D IC Technology. *IEEE Design and Test of Computers*, 22:556–564, November 2005. DOI: 10.1109/MDT.2005.134 108

[211] G. Loh. 3D-Stacked Memory Architectures for Multi-Core Processors. In *Proceedings of ISCA*, 2008. DOI: 10.1109/ISCA.2008.15 108

[212] G. Loi, B. Agrawal, N. Srivastava, S. Lin, T. Sherwood, and K. Banerjee. A Thermally-Aware Performance Analysis of Vertically Integrated (3-D) Processor-Memory Hierarchy. In *Proceedings of DAC-43*, June 2006. DOI: 10.1109/DAC.2006.229426 108

[213] X. Dong, N. Muralimanohar, N. Jouppi, R. Kaufmann, and Y. Xie. Leveraging 3D PCRAM Technologies to Reduce Checkpoint Overhead in Future Exascale Systems. In *Proceedings of SC*, 2009. DOI: 10.1145/1654059.1654117 108

[214] X. Wu, J. Li, L. Zhang, E. Speight, R. Rajamony, and Y. Xie. Hybrid Cache Architecture with Disparate Memory Technologies. In *Proceedings of ISCA*, 2009. DOI: 10.1145/1555754.1555761 108, 114

[215] G. Sun, X. Dong, Y. Xie, J. Li, and Y. Chen. A Novel Architecture of the 3D Stacked MRAM L2 Cache for CMPs. In *Proceedings of HPCA*, 2009. DOI: 10.1109/HPCA.2009.4798259 108, 114

[216] B. Zhao, Y. Du, Y. Zhang, and J. Yang. Variation-Tolerant Non-Uniform 3D Cache Management in Die Stacked Multicore Processor. In *Proceedings of MICRO*, 2009. DOI: 10.1145/1669112.1669141 108

[217] Y. Xu, Y. Du, B. Zhao, X. Zhou, Y. Zhang, and J. Yang. A Low-Radix and Low-Diameter 3D Interconnection Network Design. In *Proceedings of HPCA*, 2009. DOI: 10.1109/HPCA.2009.4798234 108

[218] K. Puttaswamy and G. Loh. Implementing Caches in a 3D Technology for High Performance Processors. In *Proceedings of ICCD*, October 2005. DOI: 10.1109/ICCD.2005.65 108

[219] J. Poulton, R. Palmer, A. M. Fuller, T. Greer, J. Eyles, W. J. Dally, and M. Horowitz. A 14mW 6.25-Gb/s Transceiver in 90nm CMOS. In *IEEE Journal of Solid State Circuits*, 2009. DOI: 10.1109/JSSC.2007.908692 110

[220] IBM. IBM Unveils New POWER7 Systems To Manage Increasingly Data-Intensive Services. http://www-03.ibm.com/press/us/en/pressrelease/29315.wss. 111, 113

[221] W. K. Luk and R. H. Denard. A Novel Dynamic Memory Cell with Internal Voltage Gain. In *IEEE Journal of Solid State Circuits*, 2005. DOI: 10.1109/JSSC.2004.842854 111

[222] X. Liang, R. Canal, G. Wei, and D. Brooks. Process Variation Tolerant 3T1D-Based Cache Architectures. In *Proceedings of MICRO*, 2007. DOI: 10.1109/MICRO.2007.40 112

[223] C. Wilkerson, A.R. Alameldeen, Z. Chishti, W. Wu, D. Somasekhar, and S.-L Lu. Reducing Cache Power with Low-Cost, Multi-Bit Error Correcting Codes. In *Proceedings of ISCA*, 2010. DOI: 10.1145/1816038.1815973 113

[224] X. Dong, Y. Xie, N. Muralimanohar, and N. Jouppi. Simple but Effective Heterogeneous Main Memory with On-Chip Memory Controller Support. In *Proceedings of SC*, 2010. DOI: 10.1109/SC.2010.50 116

[225] J. Stuecheli, D. Kaseridis, D. Daly, H. Hunter, and L. John. The Virtual Write Queue: Coordinating DRAM and Last-Level Cache Policies. In *Proceedings of ISCA*, 2010. DOI: 10.1145/1816038.1815972 117

[226] S. Liu, B. Leung, A. Neckar, S. Memik, G. Memik, and N. Hardavellas. Hardware/Software Techniques for DRAM Thermal Management. In *Proceedings of HPCA*, 2011. DOI: 10.1109/HPCA.2011.5749756 118

Authors' Biographies

RAJEEV BALASUBRAMONIAN

Rajeev Balasubramonian is an Associate Professor at the School of Computing, University of Utah. He received his B.Tech in Computer Science and Engineering from the Indian Institute of Technology, Bombay in 1998. He received his MS (2000) and Ph.D. (2003) degrees from the University of Rochester. His primary research areas include memory hierarchies and on-chip interconnects. Prof. Balasubramonian is a recipient of the NSF CAREER award and a teaching award from the School of Computing. He has co-authored papers that have been selected as IEEE Micro Top Picks (2007 and 2010) and that have received best paper awards (HiPC'09 and PACT'10).

NORMAN P. JOUPPI

Norman P. Jouppi is an HP Senior Fellow and Director of the Intelligent Infrastructure Lab at HP Labs. He is known for his innovations in computer memory systems, including stream prefetch buffers, victim caching, multi-level exclusive caching and development of the CACTI tool for modeling cache timing, area, and power. He has also been the principal architect and lead designer of several microprocessors, contributed to the architecture and design of graphics accelerators, and extensively researched video, audio, and physical telepresence. Jouppi received his Ph.D. in electrical engineering from Stanford University in 1984, where he was one of the principal architects and designers of the MIPS microprocessor, as well as a developer of techniques for CMOS VLSI timing verification. He currently serves as past chair of ACM SIGARCH and is a member of the Computing Research Association (CRA) board. He is on the editorial board of Communications of the ACM and IEEE Micro. He is a Fellow of the ACM and the IEEE, and holds more than 50 U.S. patents. He has published over 100 technical papers, with several best paper awards and one Symposium on Computer Architecture (ISCA) Influential Paper Award.

NAVEEN MURALIMANOHAR

Naveen Muralimanohar is a senior researcher in the Intelligent Infrastructure Lab at HP Labs. His research focuses on designing reliable and efficient memory hierarchies and communication fabrics for high performance systems. He has published several influential papers on on-chip caches, including a best paper award and an IEEE Micro Top Pick for his work on large cache models with CACTI. He received his Ph.D. in computer science from the University of Utah and B.E in electrical engineering from the University of Madras.